THE BOOK OF MAN

The Quest to Discover
Our Genetic Heritage

Walter Bodmer
and
Robin McKie

LITTLE, BROWN AND COMPANY

A *Little, Brown* Book

First published in Great Britain in 1994
by Little, Brown and Company

Copyright © Walter Bodmer and Robin McKie 1994
Line illustrations copyright © Mic Rolph 1994

The moral right of the authors has been asserted.

A CIP catalogue record for this book
is available from the British Library.

ISBN 0 316 90520 8

Typeset by M Rules
Printed and bound in Great Britain by
Clays Ltd, St Ives plc

Little, Brown and Company (UK) Limited
Brettenham House
Lancaster Place
London WC2E 7EN

Contents

Preface

This book tells the story of one of mankind's greatest odysseys. It is a quest that is leading to a new understanding of what it means to be a human being, and is now being carried out under the auspices of the Human Genome Project, the massive international research effort aimed at delineating the exact molecular composition of the genes that make up Homo sapiens. It is biology's answer to the Apollo Space Programme, yet it would have seemed audacious – if not downright preposterous – if it had been proposed only a decade ago.

Our aim is to help readers appreciate the scientific challenges that have been overcome in bringing genetics to this remarkable state of preparedness, and to describe the awkward problems which still lie ahead. As authors, we are concerned not with the intrigues and politics of this great scheme (though Chapter Twelve touches on them), but with the conquest of ignorance which has accompanied its development, and which will transform medical practice in the next century.

Our task would seem to be an innocuous one. Nevertheless, we are aware that controversies will surround any popular book on genetics. This preface has therefore been written to address these concerns lest there be any misunderstanding about the pages that follow.

Firstly, there is the simple question of title. In selecting *The Book of Man* we can obviously be accused of sexism. How can we explain a choice that would seem to exclude 50 per cent of the human race? Could we not have called it *The Book of Woman* or *The Encyclopedia of Humanity*? These are fair questions on which we have spent much time in debate between ourselves, and with colleagues, both female and male. In the end, we stuck with *The Book of Man* because we encountered no strong opposition to our title and because none of the alternatives had its poetic brevity, nor the requisite feelings of completeness and of relevance to our species.

More to the point, modern dictionaries provide definitions of 'man' that are entirely consistent with the subject of our book – the human race. Here is part of the definition from *Chambers Twentieth Century Dictionary*, for example. 'Man: a human being, mankind.' And for mankind: 'the human race, the mass of human beings'. Then there is the *Collins English Dictionary*. It gives a definition of man as 'a human being regardless of sex or age, considered as a representative of mankind, a person'. And finally, there is the *Shorter Oxford English Dictionary*, which gives the following definition of man: 'the human creature regarded abstractedly, hence the human race or species, mankind'.

We hope, then, that *The Book of Man* will be appreciated for its proper resonances. After all, it was Alexander Pope who wrote:

> Know then thyself, presume not God to scan,
> The proper study of mankind is man.

So if it was good enough for Pope, then surely it is good enough for Bodmer and McKie.

However, there is another accusation that is often levelled at books on genetics. This is the charge of biological determinism, an indictment which claims that modern molecular biology views human beings as lumbering robots, created and entirely motivated by their genes.

Such allegations are usually based on popular interpretations of research projects and papers. A linkage with a piece of genetic material with a human feature – cancer propensity, heart defect, homosexuality, or whatever – is presented to the public as evidence of a defective gene which will shortly be isolated and corrected, despite the lack of such claims by the actual researchers involved. For their pains scientists are then accused of assuming that human beings are entirely driven and controlled by our genes. These charges are usually unjust and we have made great efforts to ensure that they should not be levelled at this book. As authors we accept that both the environment and the chance circumstances of foetal development influence our characters and constitutions. The great prospect offered by the Human Genome Project is that by learning about how our genes affect our bodies and minds, we can subtract that influence from our equations and learn more about the others.

It is an ambitious undertaking and has been long in gestation. The conception of *The Book of Man* can be traced back to 1984 when Walter Bodmer gave the Royal Institution Christmas Lecture on the subject of 'The message of the genes'. A few years later, Robin McKie approached Walter about a separate book, and it was then the idea of the collaboration took root, eventually reaching fruition in the pages of *The Book of Man*, a book that is aimed at exploiting the disparate talents of a scientist and a journalist. The former should bring his ideas and knowledge, the latter should provide the urge for a good story and a capacity to write it. We hope readers will feel, as we do, that the combination has been synergistic and effective.

However, we would not have managed on our own, and a host of scientists gave invaluable help in the preparation of this book, both in terms of interviews and also for reading and correcting manuscripts. Our thanks therefore go to George Porter for having asked Walter to give the Royal Institution Lecture in the first place, and to Julia Allard, Edward Blake, Thomas Bouchard, David Brock, John Clegg, Kenneth Culver, Kay Davies, Paul Debenham, Peter Goodfellow, John Hardy, Erika Hagelberg, Rosalind Harding, Alec Jeffreys, Ken Kidd, the Marquis of Lothian, Peter McGuffin, Victor McKusick, Peter Martin, Robin Murray, Svante Paabo, Maryellen Ruvolo, Fred Sanger, Chris Stringer, Grant Sutherland, John Wadham and Bob Williamson.

At the ICRF, we would like to thank Janet and Yolanda for all their heroic support in translating tapes into type and disk. However, our final and greatest vote of thanks goes to our respective wives, Julia and Sarah, for their forbearance and help.

Walter Bodmer and Robin McKie
London, November 1993

1

All in the Family

Ferniehirst Castle stands on a steep slope, high on the bank of Jed Water, near Jedburgh on the Scottish–English border. It is an imposing ancient stronghold made of thick, honey-coloured Border stone and commands the Otterburn–Newcastle road, one of the main north–south thoroughfares between the two nations. Home of the powerful Kerr family for more than five hundred years, the castle was built, not for show, but for serious martial business. Its strategic importance is obvious, and its history correspondingly bloody. The Battle of Ancrum Moor was fought nearby in 1545, for example, with the Scots eventually vanquishing the English when the Kerrs switched to their side at the last minute (hence the family motto: *Sed sero Serio* – Late but in Earnest). Four years later the castle itself, which had fallen into English hands, was the scene of an even grimmer encounter. Sir John Kerr, in retaking his castle, put the entire English garrison to the sword, and then used their severed heads to play handball – a game still commemorated as the Jedburgh Ba' Game, fortunately now played with leather substitutes. The story of Sir John's gory repossession of Ferniehirst is recorded in 'The Reprisal', by Walter Laidlaw.

> So well the Kerrs their left-hands ply,
> The dead and dying round them lie,
> The castle gained, the battle won,
> Revenge and slaughter are begun.

It may not be the greatest piece of poetry ever written, but the first line is striking for one feature – its reference to the Kerrs' left-handedness. Nor is this a casual association. According to contemporary records, it is a

1

Ferniehirst Castle.

distinctive family characteristic – to such an extent that the Kerr name has become synonymous with left-handedness. Throughout Scotland, the expression Kerr-handed, or kerry- or corry-fisted, is commonly used to mean left-handed.

But it is within the confines of the castle that the Kerrs' idiosyncrasies take on a fascinating physical actuality. Unlike most abodes, Ferniehirst was said to be laid out for left-handers to live in, as were several other Kerr houses nearby. Today, the only manifestation of this distinctive design is its stair-cases. In most castles, staircases spiral clockwise. Ferniehirst has anti-clockwise ones, providing left-handed swordsmen with a distinct advan-tage, their bends giving a defender's left hand freedom to move over the open railing. This advantage may help explain another poetic tribute to the fam-ily, from James Hogg's 'The Raid of the Kerrs'.

> But the Kerrs were aye the deadliest foes
> That e'er to Englishmen were known,
> For they were all bred left-handed men,
> And fence against them there was none.

The story of the Kerr family has an irresistible appeal. It is steeped in history, revenge, bloodshed and, to cap it all, has a very clear suggestion of a dynastic legacy, in this case a profound tendency to use the left hand. We can see how this attribute has been manifested. But how was it acquired and passed on over the centuries?

To most observers the answer seems quite obvious. The Kerrs provide dramatic evidence that a gene, one of the fundamental units of biological inheritance, is responsible for the left-handedness that has marked their family like a brand, endowing them with a prowess that has had a disproportionate impact, not just on individual members or the entire family, but on the histories of England and Scotland. It is scarcely surprising then that many authorities have taken the Kerrs' left-handedness as a very serious business. In 1974, for example, the *Journal of the Royal College of General Practitioners* launched a survey, asking doctors to note the handedness of any patient bearing the surname Kerr and to send details to the journal.

The results seemed to be unambiguous. A total of 29.5 per cent of the Kerrs were reported, by both British and North American doctors, to be 'left-handed or ambidextrous' compared with only 11 per cent of a control family. No matter where a person was raised, the mere fact that his or her name was Kerr increased their chances of being left-handed almost threefold. Some predisposition must be lurking in the background, surely. Could this be the action of a gene, or possibly a cluster of genes?

However, we should take care before arriving at such a conclusion, tempting though it may be. As Professor Stanley Coren of British Columbia University, an expert on left-handedness, has pointed out, those statistics actually show that even among the Kerr family there is a strong majority of right-handers. In the general population, the survey found that 9 out of 10 individuals are right-handed. Among Kerrs, the fraction was still 7 out of 10.

But what about the evidence of Ferniehirst Castle? There is the great anti-clockwise staircase, after all, and those poems dedicated to Kerr swordsmen. Do these not suggest the manifestation of some deep biological influence? 'Possibly' is about the strongest answer that can be given to that question, for a closer look at the family's history shows that other factors apart from genes must have been involved in acquiring its distinguishing trait. Andrew Kerr, founder of the family's Ferniehirst dynasty in 1457, was certainly left-handed and found the characteristic a powerful asset in battle. His sword strokes, according to legend, swept aside those of his enemies, just as many champion tennis players and boxers disconcert their opponents with left-handed play today. But Andrew did not sit back and rely on nature similarly to endow his offspring. He specifically taught his sons and armed men-servants (who, by custom, took the family name) to wield sword and axe with the left hand. And they, in turn, did the same with their sons. We can therefore see that left-handedness, and also the family name, appears to have been acquired as often as it was inherited among the Kerrs.

And then there are the staircases. It turns out that Ferniehirst actually has five. Three are 'left-handed', two are normal. Again this seems to weaken the argument that a gene may be biasing events. However, as the Marquis of Lothian, the present owner of Ferniehirst and a direct descendant of the Kerrs, points out, the two right-handed staircases were probably used only by domestics to service the upper levels of the castle. On the other hand, the left-hand ones all open out to main doors, the crucial defensive points of the building – and that, of course, is where the Kerrs' sinister swordsmanship could be used to its best advantage.

In short, the picture is confused. Although at first glance it appears that the Kerrs' left-handedness is a straightforward biological legacy, the matter looks much more complicated on closer examination. It may be nature, or nurture, or most probably a combination of both that has so singularly marked the Kerrs. And we should not be that surprised at this turmoil. Attempts to separate the impact of our genes from the influences of our daily environment have always been awkward affairs and have constantly bedeviled studies of human attributes. In the case of the Kerrs we can say no more than that some genetic predisposition towards left-handedness probably lurks within their cells. Or perhaps it is the other way round. Human beings may be biologically programmed to be right-handed, but in some people there may be a genetic failure to achieve this status. When this happens, preference for a particular hand becomes a matter of chance. Some take the right, and others the left. As a result, left-handedness will follow through families irregularly, making it extremely difficult to track down the trait in a precise manner.

The story of the Kerrs shows how hard it can be to separate the effects of our genes from the sensory avalanche of day-to-day experience. Even without the influence of the environment, most human characteristics are complicated, multi-gene affairs that appear in varying strengths over generations, strong in an aunt, barely recognisable in a nephew and of medium strength in his daughter. Consider another human attribute: musical ability. To anyone possessing some talent with an instrument or voice, it is as natural as speaking to be able to produce fine music. On the other hand, to the tone deaf (or to be precise, the 'tune deaf'), it is an utterly baffling business. From a series of apparently meaningless blots on lined paper, some individuals are capable of making noises and rhythms that have an enormous emotional resonance. In between there are all varieties of proficiency, though at its extreme, the purest, highest form of musical talent appears like an incomprehensible, heavenly gift. How could Beethoven have been able to write those beautiful piano sonatas at the age of eleven? How could fourteen-year-old Mozart have remembered an entire mass, Allegri's Miserere, at one hearing in the Sistine Chapel and have transcribed it note for note later on? There are no simple answers to such questions. Musical aptitude, like the Kerrs' left-handedness, is inextricably bound up both with environmental influences and a person's genetic heritage. Innately gifted parents provide perfect environments for engendering prowess in their children, after all.

Nevertheless, genes must play a crucial role. The eruption of genius like Mozart's cannot be explained merely by noting he had a musician father who trained him well. Extraordinary latent talent had to be present in the first place. Equally, though, a single gene cannot be involved, for musical ability would then form a regular, cosy pattern which would be seen to run nearly through families. This simply does not happen (despite the example of the Bachs). The picture is enormously complex. Memory, manual dexterity, good hearing and a host of other attributes are involved, with as many as twenty to thirty of these characteristics sliding and coalescing over generations, striking only a very few with the full-blown mental orchestra that makes up a musical genius. Unravelling this conglomeration sounds impossible, yet progress is being made – for example, with the characteristic of perfect pitch, the ability to identify and reproduce exactly any desired note without comparing it to another. It is a power which a composer uses to summon up the internal musical imagery that is a critical part of his or her creativity. Studies by Joe Profito, a professor of psychiatry at the University of California in Los Angeles and a noted musician, suggest that perfect pitch appears to run through generations in a simple fashion. If a parent possesses it, there is a 50-50 chance that an offspring will do so. Such studies are in their infancy, of course, but they offer the prospect of one day teasing out, one by one, the individual strands that can combine to make musical genius, or any complex human quality.

Tracking the passage of inherited attributes is clearly a tantalizing, frustrating business and it sometimes seems a wonder that mankind was ever able to unravel the behaviour of genes at all. Nevertheless, we have – although their existence and rules of operation were not identified until the end of last century, by the Moravian monk Gregor Mendel. Crucially, he succeeded in this task by selecting only a few basic traits, in plants. By studying these Mendel was able to uncover the secret of the gene. By concentrating on only one or two elementary, observable attributes, he derived laws that govern the principles of biological inheritance. As we shall see in the next chapter, these rules are really very simple. That is not the problem with genetics, the study of inheritance. It is the number of genes we possess that causes the headaches. We have between 50,000 and 100,000 individual genes which control all aspects of human development from conception to

GROUP I Families with one person possessing perfect pitch

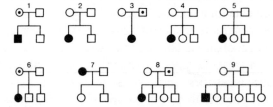

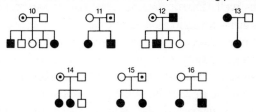

GROUP II Families with two relatives possessing perfect pitch

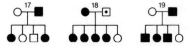

GROUP III Families with three or more relatives possessing perfect pitch

The inheritance of musical ability. Solid black circles and squares designate people with perfect pitch. Open symbols signify individuals who are known not to possess this ability, while a dot within a square or circle indicates uncertainty about their prowess.
(This chart displays a pedigree in standard form: women are represented by circles and men by squares. Individuals directly connected by horizontal lines mate, and their offspring are indicated at the end of short vertical lines.)

death, and they include our growth, stature and appearance. They also combine in complex ways that play critical roles in determining our intelligence, personality and other broad characteristics. Small wonder we can get confused.

Nevertheless, one by one, individual genes have succumbed to the attentions of modern geneticists. They have isolated and studied those responsible for the manufacture of haemoglobin, the chemical which carries life-sustaining oxygen round our bodies; insulin, which controls the breakdown of glucose and provides energy for our cells; human growth hormone which sets the limits of our stature, and many more. And in this book, we shall examine the powerful technological tools which modern geneticists have used to do this. However, we should not forget the most important source of information available to the geneticist: people, both as individuals and as entire races of human beings. Most of all, though, it is at the level of the family that the greatest amount of information, and fascination, is generated. The observation that a trait or an illness appears in some regular pattern that flows from great-grandparents down to newborn infants provides particularly important data for a scientist. The family tree – be it an ancient pedigree like the Kerrs', or a more modern affair – is a very powerful tool.

And, of course, of all the families that we could study, our own is usually the one that interests us most. So try this experiment to test your own genetic lineage. Next time you look at yourself in the mirror stick your tongue out and see whether you can make it curl, not back to front, but along the sides. If you can, there is at least a 50-50 chance that your children will be able to as well. That evens chance is the same of a newborn baby being a boy or a girl – revealing a clear-cut motif that applies to many characteristics found unequivocally in some members of a family, but not others. It is the same pattern that is laid down by the gene for perfect pitch which Joe Profito is analysing. The basis for this fundamental pattern of inheritance will be explained in the next chapter.

Or consider your sense of taste. This is clearly influenced by your heredity and at least one aspect is well defined. About 25 to 30 per cent of Europeans simply cannot taste the chemical phenylthiocarbomide, or PTC. To the rest, it has an unpleasant, very bitter tang. The pattern of inheritance in families is just like that for tongue rolling. In other words, if neither parent can taste PTC then neither can any of their children. But if one parent can and one cannot, then on average half their children will be able to taste it. And if both can taste, then so will all their offspring. Intriguingly, the ability to detect PTC is not purely a human attribute. Chimpanzees, fed with it, will frequently spit it back at researchers. Just like humans, there are chimp tasters and non-tasters, and in about the same proportions, with the former outnumbering the latter three to one. Clearly, at least one inherited disposition stretches back several million years to the time when chimps and Homo sapiens shared a common evolutionary ancestor. Beat that for a distinguished pedigree.

But perhaps the most striking demonstration of genes' influence on our physiques is provided by identical twins – the only individuals on Earth who share exactly the same complement of genes. The result is, in every case, the creation of biological doppelgängers, pairs of human beings who have precisely the same appearance, eye colour, blood groups, temperament, and a host of other features. That confluence of characteristics reveals how our make-up is laid down by our genetic heritage.

Watching genes declare themselves is obviously an absorbing business, regardless of whether you are a scientist studying bacteria or fruit flies, or a new parent or grandparent awaiting the appearance of distinctive family traits, such as red hair or blue eyes, in a newborn child. However, there are occasions when the eruption of an inherited characteristic brings only misery. This happens with genes that have gone wrong and which manifest themselves as an inherited disease. There are more than 4,000 such ailments recorded by Victor McKusick in his remarkable catalogue of genetic conditions. Most are very rare, but a few, as we shall see in later chapters, are relatively common – such as cystic fibrosis, a wasting affliction of the metabolism, which occurs in about one in 2,000 births in the West, and Duchenne muscular dystrophy, a progressive weakening of the muscles, which affects one in 5,000 boys.

Obviously each case is a personal tragedy for the children and parents involved. However, the repercussions sometimes go much further, as was the case with Queen Victoria. On 7 April 1853 she gave birth to her youngest son, Leopold. The event made medical history, for at her confinement Victoria received chloroform, the anaesthetic introduced by James Young Simpson only six years earlier. However, there was a far more profound and disturbing aspect to the birth, for it quickly became apparent that little Leopold was a very sick child. He produced cuts and scratches which would not stop bleeding. Doctors soon pinpointed the problem. Leopold had haemophilia, an inherited disease caused by a single defect in the complex physiological equipment which initiates the clotting of blood when we wound ourselves. Without this mechanism we would bleed to death at the slightest injury. And that is what happened to Leopold. He died aged thirty in March 1884, when he fell, gave himself a minor blow to the head, and suffered a brain haemorrhage. Unfortunately the matter did not end there. Two of Queen Victoria's daughters, Princesses Alice and Beatrice, passed the disease into several European royal families, including those of Spain and Russia. As Queen Victoria put it: 'Our poor family seems persecuted by this disease, the worst I know.'

Princess Alice's lineage was to suffer the greatest misery. One of her daughters, Alix, married Nicholas II, the last Tsar of Russia, and they produced one son, Alexei, who was born with haemophilia and who began to suffer from uncontrollable bleeding. In their desperation, Alexandra and Nicholas turned to the Siberian mystic Rasputin who had arrived in St Petersburg in 1903 claiming to have spiritual healing powers. He appeared to be able to help Alexei and gained considerable influence at court in the process. Slowly Rasputin gathered more and more power, until by 1915 he was taking a role in the selection of cabinet ministers and making disastrous military decisions, interference that cost the Russian army dearly in its war against Germany. Rasputin's public political meddling and his private licentious behaviour, ignored by the Tsar because he seemed able to sooth his son's terrible bouts of bleeding, helped to discredit the Russian monarchy. Eventually, a group of nobles assassinated Rasputin in December 1916, though it was

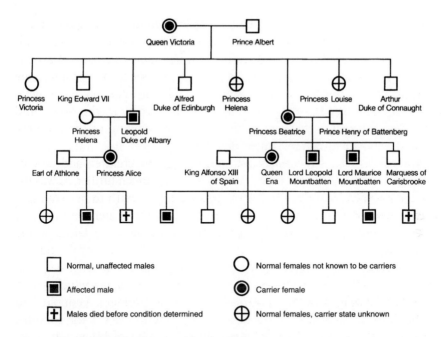

Queen Victoria and part of the royal haemophilia pedigree.

too late to save the Tsar or his family, including young Alexei. Revolution swept away their lives a few months later.

We can see Queen Victoria had a profound impact on Europe's history in more ways than one. Yet the pattern of her condition's inheritance had been recognized centuries earlier. It was known that a woman will pass on haemophilia, on average, to half her sons, while her daughters will become symptomless carriers. Indeed, a tract in the Talmud, the ancient book of Jewish law, records that a rabbi can exempt a boy from circumcision if any of his brothers have bled profusely when circumcised. Intriguingly, this exemption is extended to sons of women who have sisters with bleeding sons – showing it was known then that the disease is passed through the maternal line.

Today, the inheritance of haemophilia can be predicted thanks to our knowledge of genetic mechanisms, and its anti-clotting defect understood at a detailed chemical level. We also know the position and structure of several hundred other genes which cause inherited diseases, and have begun to unravel the chemical cascades that are set in motion by these altered biological 'triggers'.

But simple inherited ailments, each caused by single mutated genes, do not represent a major burden to the world's health, although they are obviously tragedies in their own right. What usually concern doctors, at least in industrialized nations, are conditions such as heart disease, immune ailments like rheumatoid arthritis, mental illness and cancer. They are the principal takers of lives, and some of the main causes of misery, in the West. And these conditions are also beginning to reveal their biological secrets, raising hopes that we may soon be able to tackle their pernicious symptoms, or better still their causes.

Take a relatively simple example, rheumatoid arthritis; a painful immune disorder in which the body's own defences turn on tissues in the joints and leave victims crippled, sometimes quite early in life. The condition affects about 1 in 20 women by the time they are sixty-five. Today, thanks to the techniques of modern molecular biology, we know that many early victims of rheumatoid arthritis tend to have a particular genetic make-up which can be revealed by a blood test. This has not yet produced a cure, but a start has been made.

Our basic knowledge of genetics has gone beyond medical diagnoses, however. Today it is making an impact on innumerable aspects of modern life. For instance, it is being exploited to uncover past human actions – migrations, conquests and settlements – and is also opening up new avenues in criminal investigations, by pinpointing the identity of tiny scraps of human tissue left at the scenes of crimes. We have even begun to put missing genes back into the bodies of patients to cure them of their conditions. The door has been opened on a new era.

And the key to this door lies with the creation of technologies for manipulating a substance called deoxyribonucleic acid – DNA, 'the most golden of molecules' as one of the discoverers of its structure, James Watson, described it. DNA is the true chemical of life, for it is the essential component from which our genes are made. In it is encoded the genetic language that controls our destinies. And an astonishingly powerful lexicon it is. Just six million millionths of a gram of DNA carries as much information as ten volumes of the *Complete Oxford English Dictionary*.

It sounds incredible, yet that single scrap of DNA is the critical product which is created when egg meets sperm, joining the inheritance of a mother and a father in equal parts. In this way, a single cell – the fertilised egg – is produced, one that has the potential to form a new and unique individual under the guidance of the DNA within its nucleus. The human body is made up of a hundred million million cells of many different sorts, and all contain the inherited information that comes from that first, single cell created at fertilization. These instructions carry the biological details necessary for making all the different tissues and organs, and the cells of blood, skin, kidneys, lungs and many others including, of course, the brain. These guidelines make each of us unique – apart, of course, from identical twins. As the

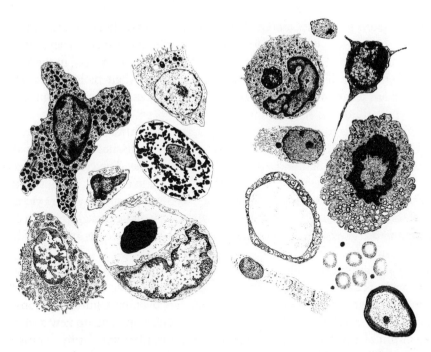

The spice of life: some of the different types of cells that make up the human body. The construction of each is directed by a single scrap of DNA that weighs a mere six million millionth of a gram.

embryo, and then the child, matures, the DNA script within its cells is read and translated into proteins from which tissue, nerve cells and hormones are constructed. These in turn are transformed into organs, thought processes, memories, and even behaviour patterns, which range from instinctive flinch reactions to complex tendencies that include our elusive musical talents, and the urge to use our left or right hands.

This is the Book of Man, the instruction set according to which all humans are made. And when we learn how to read its pages and chapters we will have obtained information relevant to the understanding of most diseases, individual differences in behaviour, and a new awareness of our own history and evolution. Our aim in this book is to describe how the Book of Man is now being read and interpreted.

We shall begin, in the next chapter, with the discoveries that were initiated by Mendel, and which led to the uncovering of the existence and the behaviour of the gene, before we go on to explore the structure of its constituent molecule: DNA. Then we shall look at the dramatic story of the Lords of the Genome, those men and women who first learned how to manipulate DNA and who generated a whole new industry, biotechnology, in the process. In

Chapter Five, we shall look at the first beneficiaries of this new science – the victims of inherited illnesses, conditions that have blighted generations but which we are now learning how to halt in their tracks. Then in Chapters Six and Seven, we shall look at illnesses – cancer, diabetes, and others – which are far more common and which are also succumbing to the molecular biologist. The mind, the subject of Chapter Eight, is also revealing its genetic secrets in just the same way as our history as a species is being revolutionized by the science of genetic archaeology, the topic of the following chapter. In Chapter Ten we shall witness the revolution in forensic science that has been wrought by molecular biology, and and in Chapter Eleven, we will look at the fledgling, but infinitely promising, technology of gene therapy in which missing genes are actually being inserted into those afflicted by inherited ailments. The book's penultimate chapter will tell the personal story of Walter's own involvement in genetics, before we close with a discussion of the enormous ethical implications of this most exciting of sciences.

This last chapter will be of crucial importance, of course, for the impact of molecular genetics may frighten some people. Will we not trigger a dangerous biological determinism by defining a person's potential at birth, they ask? And are we not at risk of playing God by creating designer human beings, made to measure, like off-the-peg suits? Without wishing to sound complacent, the answer to both questions is simple. We are very far away indeed from such a prospect, and nothing like it is in any scientist's mind. As we have seen with relatively simple concepts such as left-handedness, the genetic roots of most human attributes remain tantalizingly elusive. They will be uncovered, undoubtedly, but not perhaps for many years. As for complex measures such as personality, we are many, many decades away from defining them genetically. In the meantime, pinpointing those 50,000 to 100,000 genes, and deciphering their role and function is likely to bring unqualified benefits: new drugs, methods of diagnosis and methods for studying human behaviour.

And far from ruling out the role of the environment, the decoding of the Book of Man will only enhance it, by revealing its precise influence and impact. Once we have understood the function of genes in behaviour, growth and disease we can eliminate them 'from our enquiries' and reveal the subtle but equally profound effect of nurture in defining human actions. And when we have done that we may well have answers for those puzzling questions about how we acquire the characteristics that distinguish us as individuals: our musical ability, our taste in food, our choice of friends and, of course, our preference for the hand with which we write, and play sport, and which determines whether we are to be right-handed, like most people – or corrie-fisted like the Kerr dynasty of Ferniehirst Castle.

2

The Dice of Life

Mankind has had a historic penchant for fiddling with genes. In the nineteenth century, for example, practitioners of animal husbandry cultivated an expertise – which had already been a potent, driving force during the agricultural revolution – to a new level. Fresh breeds of lean cattle were developed, sheep with thick, rich fleeces were introduced to meadows and pastures, and pigs of unsurpassed fatness and fecundity brought to the farmyard. And with the development of each new strain animal breeders, particularly in England, honed their talent until it became an art.

But it never became a science. For all their achievements, livestock owners, such as Robert Bakewell who 'created' Leicestershire Longhorn sheep and various improved farm horse strains, could not work out the clear-cut scientific principles behind their work. Without rigorous rules to guide them, their powers were destined to be limited.

This concoction of success and failure was observed far away, in Moravia, in the Austro-Hungarian empire in the first half of the nineteenth century and now part of the Czech Republic. Landowners there had heard about English breeders' expertise, and sought to emulate it. However, they also wanted to improve it. They wished to harness this primitive biological engineering within a proper scientific framework, and turned to F.C. Knapp, the abbot of Brno monastery.

To the modern mind, inured to the notion that ecclesiastical and scientific minds must be divided by unbridgeable schisms, this approach may seem odd. Yet at the time it would have been viewed as a perfectly normal request. After all, as abbot of a large monastery, Knapp controlled large tracts of farm land. Also, as head of a great centre of learning, he had access to some of the country's finest minds. In particular, Knapp was able to enrol the genius of Gregor Mendel.

Gregor Mendel.

Born on 22 July 1822 in the hamlet of Heinzendorf, Gregor Mendel (he was actually christened Johann, and only assumed the name Gregor when admitted to the novitiate at Brno in 1843) was raised in a peasant family shackled by feudal poverty. His father Anton had to spend three days a week working on his landlord's land, for example. On the other hand, Anton was also ambitious and hard-working. In addition, his son received some exceptional help from the local priest, Father J. Schreiber.

Schreiber had been a pioneer teacher of natural history at an educational institute at nearby Kunin, until expelled by Jesuits for 'introducing foreign Lutheran ideas'. So Schreiber moved to Heinzendorf, establishing a fruit-tree nursery from which he taught parishioners to cultivate and graft different varieties. One recipient of these grafts was the Mendel family.

However, Schreiber did more than nurture the family's love of horticulture. He also recognized the quality of their only son's intellect. More importantly, he acted upon this realization, persuading the Mendels to send the boy to a succession of good schools. Young Mendel thrived there, though poverty would have ended his studies had not his teachers persuaded Abbot Knapp to admit the young student to Brno. So, aged twenty-one, Mendel became a cleric, so he could free himself 'from the bitter struggle for existence', as he later put it.

At Brno, Mendel was liked for his humility, intelligence and competent teaching. However, he failed his teacher's examinations – ironically, because of his weakness at biology. So Knapp sent him to Vienna University where he studied a host of subjects including physics under Professor Christian

Doppler (of Doppler effect fame). He became an adept experimenter and also encountered the concept of discrete units – atoms, molecules – which were then being introduced to explain physical and chemical phenomena.

In all, it was a perfect preparation for his life's work. Intelligent, familiar with the ways of advanced horticulture, influenced by remarkable men such as Schreiber, Knapp and Doppler, and steeped in the latest thinking in physical and statistical sciences, Mendel was intellectually as well-armed as a young researcher could be. He was undeniably a very different person from the Gregor Mendel who is standardly portrayed as an obscure Moravian monk, stumbling upon the golden laws of biology while pottering in his garden. Certainly, he was an excellent choice for fulfilling Knapp's request, made on Mendel's return to Brno, that he uncover the scientific principles of breeding.

What puzzled breeders at the time was the lack of recognisable patterns or rules to inheritance. They could see characteristics being passed between generations of livestock and cereal crops – colour of blossom, an animal's growth, or number of grains in an ear of corn. But the process seemed haphazard. Sometimes traits missed a generation, sometimes they simply disappeared.

To cut through the confusion, Mendel realized it was vital first to isolate clearly defined plant characteristics. Instead of trying to account for all their various traits at once, he selected only one or two very clear properties. He also decided that the more separate measurements he made the more likely it was that a mere chance effect would be eliminated. In doing so, Mendel pioneered the application of statistics to biological research.

But first he had to spend several years merely establishing the true breeding nature of his choice of experimental subject, the common pea. In other words, he had to create pure lines of plants, separating those which only produced round seeds from those that only produced wrinkled seeds, or isolating those with only short stems from those with only long ones. Such are the labours of the true genius!

Then Mendel began his historic work, by crossing his pure breeding stocks. For instance, he crossed pea plants with round seeds with those with wrinkled seeds. This produced a first generation that had only round seeds. The inherited attribute – round seeds – was therefore said to be dominant. The wrinkled seed attribute was called recessive.

Mendel let this first generation of plants self-fertilize (a common occurrence in horticulture). And when he did so, he found that the wrinkled seeds reappeared, and in an exact ratio. This second generation, he discovered, had one wrinkled seed for every three round seeds.

But Mendel did not stop there. He let the second generation self-fertilize as well, and once again he found a clear-cut relationship: two out of three round peas always produced some recessive, wrinkled seeds on self-fertilization. Wrinkled seeds when self-fertilized would always produce only wrinkled peas.

This behaviour sounds quite baffling, with wrinkled peas disappearing and reappearing like demented jack-in-the-boxes. But Mendel devised a beautiful schematic model to explain his results. He assumed that the 'elements', or genes as we would now call them, which determined his characters, were discrete entities. And he also reasoned that a gene must be carried in a double dose in each pea. The fact that one gene was 'dominant' over the other gene, the 'recessive' one, explained why those bizarre leaps in inheritance were occurring.

Take the first generation that was produced from the two pure lines of round-seeded and wrinkled-seeded pea plants. Each of these parents contained a pair of genes that controlled its seed shape (round or wrinkled). A plant passes on one of this pair to its offspring. The other parent passes on the other gene. This means that every plant in the first generation must have been given one round gene and one wrinkled gene. And as round is dominant over wrinkled every seed should have been round – as Mendel's experiments revealed.

When we look at the second generation that was created, we can see that there are now four equally likely possible combinations of pairs of genes: two round genes, one from each parent plant; a round and a wrinkled; a wrinkled and a round; and two wrinkleds. As round is dominant, this means that in three of four cases, a plant with round seeds is produced. Only in the last case are wrinkled seeds produced. And there, in a nutshell – or a peapod – are the laws of Mendelian inheritance.

Yet they were drawn up in the absence of any notion of the mechanisms of inheritance – a remarkable achievement that establishes Mendel as the founder of the science of genetics. We now know that the ratios he worked out apply equally to humans, from the genes which are responsible for eye colour, brown (dominant) versus blue (recessive), to matings between two normal carriers of the cystic fibrosis gene. The cystic fibrosis gene is recessive and will give rise to children with two such genes (and who therefore have the disease) in a quarter of all offspring produced by carrier parents.

Fame was to elude Mendel, however. He reported his findings in the obscure journal *Verhandlungen*, and sent copies of his results to several leading biologists. None took any notice and for thirty-five years his momentous research languished in obscurity, until 1900 when it was rediscovered and its importance recognized by the botanists Hugo de Vries, Karl Correns and Erich von Tschermak. And the fact that the twentieth century dawned with

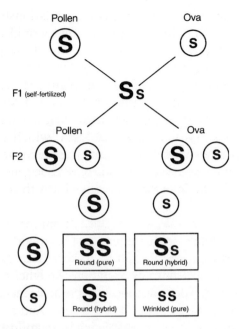

Mendel and the Pea Plants.
Mendel's first generation (F1) of pea plants was created from two pure lines of round-seeded (S) and wrinkle-seeded (s) stock. Each plant in this generation had one round-seeded gene (S) and one wrinkle-seeded (s) stock. Each plant in this generation had one round-seeded gene (S) and one wrinkle-seeded (s). As round is dominant, every seed was round. A second generation (F2) was then created from the first, and this produced four combinations: two round genes, one from each parent plant; a round and a wrinkled; a wrinkled and a round; and two wrinkleds. As round is dominant, in three out of four cases, a plant with round seeds was produced.

the uncovering of the forgotten laws of inheritance is apt. After all, ours has become a century increasingly dominated by our awareness of our biological determinacy. It should end with the unravelling of the very molecules that control that process.

The work of Gregor Mendel was not the only trigger to set loose this revolution. Other scientific forces were gathering. Just as Mendel's great works were being revealed by Dutch scientists, a young research assistant in Vienna was making discoveries that were to have an impact of equal immediacy on our awareness of heredity and health. His name was Karl Landsteiner and his interest was blood, and human blood in particular.

Human preoccupation with blood is as old as history. The ancient Egyptians advocated baths of it for recuperation; Roman gladiators believed that by drinking the blood of fallen adversaries they could acquire some of their courage and strength; blood is mentioned several hundred times in the

Bible. In short, the rejuvenating potential of blood was well understood. As Goethe put it: '*Blut ist ein ganz besondrer Saft*' (blood is a most extraordinary juice).

The trouble was in finding a method for exploiting its power scientifically. Until William Harvey discovered how blood circulated round the body, the only way to administer transfusions was by mouth. After Harvey, however, physicians realised that it could be transfused by injecting it into the blood-stream. It was the great British draughtsman, Sir Christopher Wren, architect of St Paul's Cathedral, who first used quills and silver tubes for injecting drugs and other substances into the veins of dogs and who suggested that this method could be exploited as a means of transferring blood from one individual into another.

Doctors then found that all bloods are not alike. For instance, when transfusions were attempted from one animal species to another, or from animals to humans, or from one human to another, there were inexplicable reactions. Patients suffered burning sensations in their arms, pain around their kidneys, sweating, vomiting, and diarrhoea, and later the voiding of urine as 'black as soot'.

Then, in 1875, the German scientist Landois noticed that if he took red blood cells from one animal, say a cow, and mixed them with serum from another, say a sheep, the red cells very often clumped. (Blood is a mixture of many different cell varieties, of which red cells are the most numerous. When a vein or artery is injured, blood cells will stick together and close the wound to prevent further blood loss. The pale fluid left behind is called serum.)

What Landsteiner wanted to know was what would happen if he did similar experiments using blood samples taken from humans. So using samples from his associates, laboratory helpers and others he began a series of simple experiments. What he discovered created the basis of modern blood transfusions, and a great deal more.

Landsteiner found that he could classify people into three different groups called A, B and O. Serum from O people clumped the cells of those who were A and B, but not those of another person who was O, while serum from a B type clumped the cells of someone who was A but not B or O and the serum from an A type individual clumped the cells of someone who was B but not those who were A and O. Later a fourth type was found – a combination of A and B. (The precise cause of this clumping will be explained in Chapter Seven.)

So, starting from this simple idea and a straightforward test, Landsteiner defined the first blood groups in 1900. In doing so, he laid the foundations for the development of human genetics, a fledgling science for a fledgling century.

Diagram of A, B, O, Blood Clumping

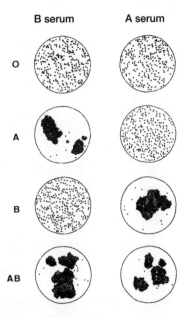

How Landsteiner discovered the grouping of blood. He found he could classify people into three different groups: A, B and O. When mixed together, O serum clumps A and B blood cells but not O cells; B serum clumps A cells but not B or O; and A serum clumps B cells but A and O. Later, when AB blood was discovered, it was found that it was clumped by all other serums. The clumping is caused by anti-bodies called anti-A and anti-B found in the serum (see Chapter Seven).

The genetic interpretation behind Landsteiner's work is simple. There are three versions of one gene, an A-determining version, a B-determining version and an O for neither. Thus O individuals had the constitution OO, AB individuals the constitution AB, while A individuals can be either AA or AO and B individuals either BB or BO. These types are inherited following exactly the laws that Mendel had worked out from his experiments with pea plants in Brno. Thus, for example, a mating of AB by OO would give one half offspring who were A (genetic formula AO) and one half offspring who were B (genetic formula BO). The idea that the genes came in pairs that were distributed to the offspring with equal chances, just like tossing a coin, seemed to be quite fundamental.

The patterns produced by this biological coin-flipping can be intriguing. In the south of England, for instance, blood type frequencies divide up in the following way: O group, 44 per cent; A, 45 per cent; B, 8 per cent; and AB, 3 per cent. However, these frequencies vary quite markedly between different populations. Type B is nearly three times as common in oriental

Inheritance of Blood Groups

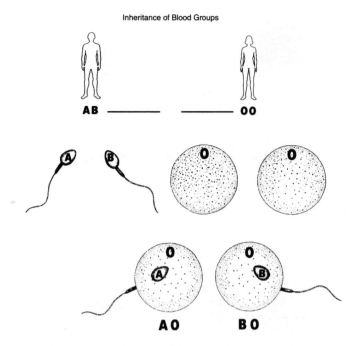

Mendel's laws and blood. A man with AB blood and a woman with O (genetic formula OO) blood would produce offspring of whom half would have A blood (genetic formula AO), and half would have B (genetic formula BO).

populations as it is in Europeans, for example. Even within Britain the frequency of type A varies quite significantly, increasing from south to north. And as we shall see in Chapter Nine, these ABO blood types were the first genetic variations to be used to study differences between populations.

Now the great advantage in studying factors such as blood groups is that you actually get down to the level of the chemistry of living material. The uncertainty associated with vague measurements such as height, or even attributes that seem clear-cut, like hair colour or baldness, begins to disappear. At the cellular level, genetics becomes a basic pursuit.

Suddenly, the mechanisms of inheritance and the existence of at least one example of their operation inside the human body had been opened up to exploration. Yet many other questions were left unanswered. In particular, scientists wanted to know much more about the passage of this inherited information. What was the means by which our genes expressed themselves? In particular, how do two daughter cells obtain exact copies of their genetic instruction sets from a parent cell? The answer came, not from observing human cells but, rather bizarrely, from studying those of the threadworm of the horse. The reason, however, was simple – the cells of the

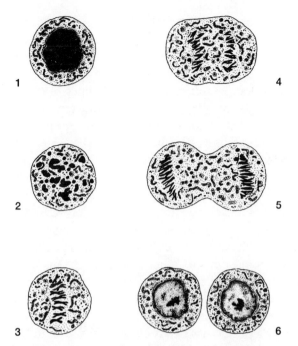

Multiplication: a cell divides and its thread-like chromosomes appear and divide, one copy going to one daughter cell, another to the second daughter cell.

threadworm are large enough to be seen easily under a microscope, so it is possible to work out what is happening during cell division. And what researchers saw was of tantalizing interest. Just before a cell divides, they could make out that thread-like structures were beginning to form. These seemed to divide along their length, with one newly created spindle going to one daughter cell, and the other to the second cell. This division is known as mitosis and it accounts for the creation of all our bodily cells from the first single cell created at conception. And the spindle structures were called chromosomes because of their power to absorb special dyes during staining for microscopy. Chromosome literally means 'coloured body'.

These observations were first made in the 1880s, and were eventually responsible for leading the great German zoologist August Weismann to propose that chromosomes are the carriers of the hereditary instructions. As he pointed out, they fulfil the precise requirement of being passed on equally from parent to daughter cells during cell division. Once again we see the evidence of the powers of genetic science gathering as the new century dawned.

However, Weismann had another contribution to make to this rapidly forming body of knowledge. He pointed out that in order to maintain constant chromosome numbers from generation to generation, the sperm and the egg would each have to be special in having half the normal number of chromosomes. In this way the fusion of an egg and a sperm which results in the creation of the first cell of the fertilized embryo would reconstitute the normal number found in all other cells in the body.

If we look at a photograph of the chromosomes of a dividing human female cell, we see a jumble of little sausages of different sizes each with particular banding patterns. It looks haphazard and meaningless. However, when we arrange these sausages carefully by size, shape and banding pattern, starting with the largest and ending with the smallest, we see that a human being's 46 chromosomes occur in pairs – 23 in total. And each pair is quite distinctive, so they can be numbered from 1 to 22 – and X.

This last X sounds intriguing and so it should, for when we carry out the same examination of a human male cell we find the same 22 pairs as we did with the female cell. However, in this case there is no pair of Xs left over. We see only one X chromosome, together with a much smaller, stubbier little chap – the Y chromosome. And that is all that distinguishes the sexes, at least as far as a geneticist is concerned. The mysterious works that divide the sexes, and the alchemy that mutually attracts them, boil down to a simple matter of whether a person has two X chromosomes, or an X and a Y.

And the crucial process involved in creating this magical difference is the process of meiotic division that was first uncovered by Weismann. When men produce sperm in the cells of their testes, each of the chromosomes of their body's cells splits up, with one member of a pair going into each sperm. (This process is called meiosis, and it contrasts with mitosis, the normal form of division which we encountered on page 20 in which two daughter cells, each bearing 46 chromosomes, are found from one parent cell.) So unlike the normal cells of the body, which will have two chromosomes 1 and two chromosomes 2 and so on, each sperm cell has just one copy of each chromosome: numbers 1, 2 and so on.

However, when it comes to X and Y, each sperm gets either an X or a Y chromosome, purely as a matter of chance – like tossing a coin to see whether it is heads or tails. And so, on average, one half of the sperm will carry an X chromosome and the other half a Y chromosome. (In a woman, of course, all her eggs will carry just one copy of the X chromosome, just as they do for all the other chromosomes.) Thus, if an X-bearing sperm meets an egg the resulting fertilized egg will be XX and female, while if a Y-bearing sperm meets an egg the resulting fertilized egg will be XY and so male. Since most men have roughly equal numbers of X and Y sperm there are equal numbers of XY and XX fertilized eggs and that is why there are roughly equal numbers of males and females. Thus it is that the father determines (unwittingly, of course) the sex of his offspring, suggesting some awkward

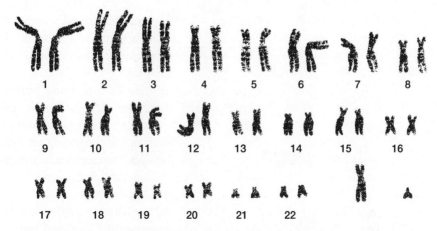

*The variety of life. The human genome is arranged into 23 pairs of chromosomes.
These are numbered from 1 to 22, with sex-determining chromosomes (on the bottom
row, far right) being designated X and Y.*

historical home truths. Take Henry VIII. His dismay at those of his queens
who did not bear him male heirs was clearly misdirected. The solution to the
Tudor male lineage lay within his own regal testicles.

Such a vision of genetic division and recombination is standard biological
fare today. But it seemed revolutionary stuff at the dawn of the twentieth century. In this case, the breakthrough was made, in 1901, by a young American
zoologist, Walter Sutton, who realized that the behaviour of chromosomes
could explain entirely Mendel's rules of inheritance. And so it was that the
abstract idea of Gregor's elements, which we now call genes, could be associated with concrete entities seen down a microscope – the chromosomes.

Nor is this process of dual division limited to the sexes. The other non-sex
chromosomes, those from 1 to 22, fall within its influence. Take the blood
types discovered by Landsteiner. We now know that the gene that determines ABO groups lies on chromosome 9. Each of us has two different
chromosomes 9, one inherited from our father and one from our mother.
And if a person gets his paternal chromosome 9 with an A blood group gene
on it, and his maternal chromosome 9 with a B version, this means he or she
will have AB blood. And if that person is female, she in turn will produce
eggs that will have either the gene responsible for A blood or the one for B
blood. In turn, if her husband has type O then he can only contribute,
through his sperm, a chromosome 9 carrying the O gene. Their children will
therefore have the genetic constitution AO, and so have type A blood, or will
have the genetic constitution BO, and will have type B blood.

Now we know there is no connection between the sex of a child and its
blood group; and that is because the sex-determining X and Y chromosomes

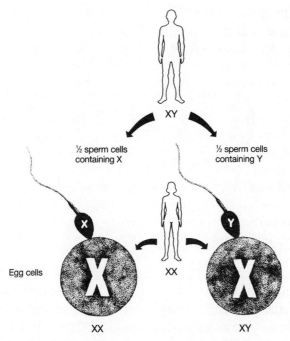

Sex determination. Each sperm cell made by a man either gets an X or a Y chromosome, purely as a matter of chance. This determines whether a child will be female or male.

are separate entities which are passed on independently from the blood group determining chromosome 9. This is true for all chromosomes. However, this is not the case for two genes that are on the same chromosome, for example haemophilia and colour blindness. They both reside on the X chromosome. So are they always inherited together? The answer is no, because of a process called recombination which occurs during meiotic division when the sperm and egg are created. This mechanism swaps sections between a chromosome pair, for example the two X chromosomes carried in a female, to create a new X chromosome in one of her eggs. In this way new chromosome combinations are created, maintaining high genetic variability in a population. However, when two genes are close together, the chances that recombination will separate them will be low. If they are far apart on a chromosome, that prospect will become almost certain. Consider a woman with both the haemophilia and the colour blindness gene on one X chromosome (the other being normal). Thanks to recombination, about 6 per cent of the eggs she produces will either have the haemophilia and a normal gene, or they will have the colour blindness gene and the normal blood-clotting gene; that 6 per cent figure is a measure of how far apart the two genes are on the chromosome. For the remaining 94 per cent of her eggs, the colour blindness gene will continue to act as a genetic marker that flags the haemophilia gene. The importance of this process for tracking genetic diseases through families will be discussed in Chapter Five.

Understanding how blood groups track through generations was one of the great triumphs of genetics in the first years of our century, for it provided knowledge of immense practical value. As scientists realized, if a woman had type A blood and her husband had type O, and if they had a child that was AB, there was only one conclusion. The husband was not the father of the child. Such a blood test does not help in proving the fatherhood of a child, of course. Any man with B blood could claim parenthood of the child purely on the strength of his blood. The technique's power lies with its ability to rule out potential fathers. And in that sense its results are unequivocal, though courts have sometimes been shamefully reluctant to appreciate this point.

In 1945, for example, Charlie Chaplin was sued for paternity by the starlet Joan Barry who had, in 1943, given birth to a daughter Carol Ann. Chaplin was the father, she claimed. Yet blood tests, carried out before trial, showed Chaplin had O blood, Joan Barry had A blood, and the baby had B blood. In straight terms, Chaplin was not the father. Yet a motion to dismiss the case against him was overruled by Judge Stanley Mosk. He claimed, enigmatically, that the 'ends of justice will best be served by a full and fair trial of the issues.' After retrial, a jury voted eleven to one that Chaplin was the father of baby Carol Ann, a biological absurdity of such breathtaking magnitude that it led, within a few years, to the state of California introducing legislation to prevent pursuance of paternity cases where blood tests had conclusively proved a defendant could not be the father of a child.

Modern genetic techniques are much more powerful and precise and are able not only to rule out someone as a father, they can also rule them in, as we shall see in Chapter Ten. However, the lesson – that the law ignores the veracity of biological principle at its peril – remains unchallenged.

Peas, threadworms and blood from laboratory workers; these were the unlikely constituents of a scientific revolution that unfolded with startling rapidity. Beginning with the rediscovery of Mendel's laws and leading us through Landsteiner, Weismann and Sutton's historic work, this witch's cauldron of ingredients was exploited to reveal the behaviour of genes, and those of human beings in particular – and all within a few years.

The next great challenge – to find out what are our genes are actually made of – was to be solved in even more remarkable circumstances. It forms the core of our next chapter.

3

DNA: Life's Mother Tongue

In 1904, a young student from Grenada in the West Indies arrived in Chicago. His name is not recorded, and the details of his life have been lost. Nevertheless, this anonymous immigrant played a strange but important role in our understanding of the Book of Man. His tale is crucial, so, to simplify our exposition, we shall call him Joe Smith.

Not long after his arrival, Joe began coughing, before developing a persistent fever. In hospital, he was examined by Dr James Herrick, whose diagnosis was simple – Joe was anaemic. But when Herrick looked at Joe's red blood cells he found the problem lay not with their absence (the usual cause of anaemia), but with their shape and size. Instead of being invariably round and uniformly stuffed with life-giving oxygen, his red cells were a hotch-potch of sizes and shapes. In particular, a large number were thin and crescent-shaped. Herrick concluded that Joe's illness must be due to intrinsic changes in his red cells and therefore not the result of an infection or any other known cause of anaemia. As Herrick put it: 'Some change in the composition of the corpuscle itself may be the determining factor.'

Herrick later presented Joe's case at the twenty-fifth annual meeting of the Association of American Physicians in Washington in 1910. Despite its intriguing content, there was no discussion. None of his distinguished audience recognized the special nature of the disorder Herrick had uncovered; a considerable oversight given the ailment's subsequent special role in the history of genetics. For, as we shall see, it became the first inherited disease to be characterized at a molecular level. Indeed, sickle cell anaemia (as the condition was later named) has turned out to be the forerunner of a host of blood abnormalities which have taught us so much about inherited disease mechanisms and about natural selection in human populations. It was not the first time that a vital breakthrough in genetics went unrecognized for decades

Herrick was an intriguing character. An English teacher turned professor of medicine, he never forgot his literary roots. His final address to the Association of American Physicians was an after-dinner talk on 'Why I Read Chaucer at Seventy', a tendency which he simply attributed to the fact that 'I read him at nineteen.' Such polymath inclinations were to become a common theme in the chemical exploration of the gene, as we shall see.

Gradually more and more cases like Joe's were discovered and sickle cell anaemia became an established disease. Herrick found that his patients had parents who, at least as far as the anaemia was concerned, appeared to be quite normal. Then a colleague, Victor Emmel (this time, a Kansas farmer turned anatomist!) made a fascinating discovery: when he removed blood from a patient's non-anaemic father, he found he could observe those funny sickle-shaped cells that Herrick had discovered in full patients. Parents, they realized, might be clinically normal, but they still had some sickle cells. Clearly the trait was somehow being passed on from generation to generation.

However, it took another twenty-five years before scientists realized exactly how – when James Neel, a Michigan University physician, worked out the absurdly simple explanation. The 'normal' parents with the sickling trait were carriers of a defective gene whose damaging effect was offset by the presence of a corresponding normal gene. Only when two such sickle-trait carriers mated did they produce offspring of whom one-quarter would inherit the abnormal sickle gene from both parents and so have the full-blown disease. It was the classic behaviour of a recessive illness – one that exactly fitted Mendel's rules of inheritance.

Now there was one crucial feature about Joe Smith, in whom James Herrick first detected the sickle cell trait. He was black. And this subsequently proved to be no mere chance phenomenon. Population surveys of American black people showed that a relatively high percentage were sickle-trait carriers, while none was found among people of European background – which suggested that the gene had been brought to America by African slaves, the originators of America's black population. And true enough, sickle cell anaemia was eventually discovered in Africa in 1925, when a ten-year-old Arab boy was admitted to hospital in Omdurman, in Sudan, with a high temperature and fever. His symptoms suggested malaria, and so his physician, Dr Gwendolyn Hampton, examined his blood for malarial parasites – and found sickle-shaped cells instead. Far from being rare, the new disorder turned out to be ubiquitous.

Over the next twenty years, more and more cases of the disease were

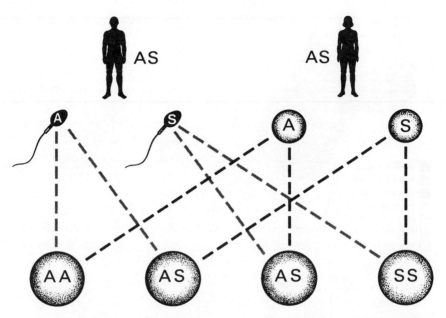

Sickle cell inheritance. A carrier father has two genes, one for sickle haemoglobin (S) and one for normal haemoglobin (A). A carrier mother has the same pattern. On average these parents will produce one child with no sickle gene at all; two who are carriers with only one sickle gene each; and one who has two sickle genes and who manifests the disease.

uncovered and its global prevalence established. However, there was one statistic that puzzled scientists. They found that full-blown sickle cell anaemia was less common among African black people than among American black people. Yet in some places – the Gambia, the Gold Coast, Nigeria and Cameroon – nearly 20 per cent of the population carried the sickle cell trait, three times the frequency found in American black people. So where were all the African anaemia cases? The answer soon became clear. Nearly all were dying before they were three years old. Only better living conditions in America were keeping alive its sickle cell population. But if that were the case, without the advanced care of early twentieth-century Western medicine, the sickle cell gene should have been eliminated long ago in Africa. In killing off so many young victims with double doses, the gene should surely have eradicated itself. What could be sustaining this catastrophic gene in African populations? The answer, we now know, is malaria.

Doctors have discovered that there are significantly fewer cases of malaria among sickle-trait carriers than in the rest of the population. In addition, those parts of Africa where malaria is rife exactly match those where the frequency of the sickle cell trait is highest. Stick one distribution map over the other, and the fit is remarkable, though, as we shall see in Chapter Nine,

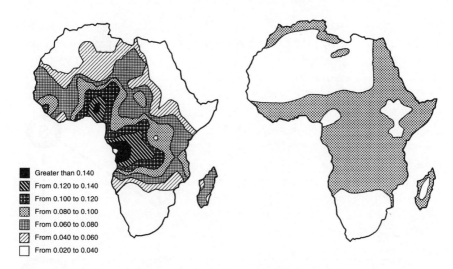

Greater than 0.140
From 0.120 to 0.140
From 0.100 to 0.120
From 0.080 to 0.100
From 0.060 to 0.080
From 0.040 to 0.060
From 0.020 to 0.040

The geography of sickle cell anaemia. The map (left) shows the frequency of the sickle-cell gene in populations in Africa, ranging from areas in the north and south where it is uncommon (unshaded in map), through shaded zones, until we reach regions (totally inked-in) where it is at its most frequent. This map closely corresponds to one of malaria's distribution in Africa (right).

other inherited blood disorders also thrive within the penumbra of malaria. In the case of sickle cell anaemia the conclusion is inescapable. Carriers are more resistant to malaria and so live longer to produce more offspring – who of course carry the sickle cell trait. Only when two carriers meet, marry and produce children does this 'arrangement' go catastrophically wrong.

It is not obvious why these illnesses should shadow each other in this way, though there are pointers. Malaria parasites grow inside red blood cells, which implies that carriers' cells do not provide such a congenial environment as those of 'normal' people. Nature therefore seems to be striking a cruel bargain with humanity. Young sickle cell anaemia victims are paying for the advantage that their carrier parents have accrued in being more resistant to malaria than the 'normal' person.

Malaria is probably relatively new as a major disease for mankind, arriving in its current intense form with the development of agriculture about three thousand years ago. When forests were cleared and irrigation channels created, small pools of stagnant water formed, providing ideal homes for mosquito larvae that carry and transmit the malarial parasite. And as farming spread, malaria would have become more and more troublesome. Sickle-trait carriers would therefore have better survived malarial attacks, and increased in numbers at the expense of their 'normal' relatives. And that, quite simply, is how Darwinian natural selection operates at a human level.

This means that when black Africans were brought to malaria-free America as slaves, many must have carried with them the now useless sickle gene as an added burden to their suffering.

The sickle cell story is just one example of how gene geography can help to define and elaborate the history of human migrations and population mixtures, a topic to which we shall return in Chapter Nine. And it is a story that sheds light elsewhere – for, crucially, it starts with a mutation in a gene. The question was: what is a gene and how, when it is altered, does it give rise to abnormalities like sickle blood?

To find the answer, researchers had to turn not to general issues, such as the growth patterns of peas or the clumping tendencies of human blood, but to the very basic chemistry of living tissue. These two approaches – the specific and the general – would coalesce, as we shall see, with striking fruitfulness later in the century.

If you could take a piece of tissue, human or otherwise, and break it down into its smallest chemical constituents, you would end up with a large mound of atoms, mainly those of the elements hydrogen, carbon, oxygen and nitrogen. But of course chemicals are not just amorphous piles of elements, they are made of molecules, which are particular combinations of elements. So if we really want to analyse the chemistry of living material we have to ask: what molecules do you find in a living cell?

Consider what happens when you fry a piece of bacon. Fat melts off and you are left with cooked meat. The fat is made of hydrocarbons – strings of carbon and hydrogen atoms – while the meat is essentially protein, a complex form of molecule that is the essential building block of living organisms and to which our brief history of genetics must now turn.

There are thousands of different types of proteins. Our skin is made of types called keratin and collagen; the lens of the eye is constructed out of crystallin, which gives it its clear and flexible structure; egg white consists mainly of albumen (which is also found in blood); and the major constituent of milk is a protein called casein. In addition, some proteins play active, not just structural, roles within our bodies. These are called enzymes and they are the working molecules of life. They degrade our food into smaller molecules which can then be used to generate energy, and they create new chemical structures for cells as they grow and divide.

And it is this category of protein, the enzyme, that helped place the next piece in our genetic jigsaw. In 1902, the British physician Archibald Garrod came across a patient with severe back pain, early onset arthritis, and ears and nose coloured an odd, blueish-blackish hue, as if bruised. Even his sweat was darkly coloured. But most striking of all was Garrod's discovery that his patient's urine would turn black if left to stand for some time.

Having recognized one case of this strange disease, Garrod sought others and found that, although rare, the condition appeared to run in families. After consulting with William Bateson, the first real British geneticist, Garrod realized that the disease, which he called alkaptonuria, was a classic recessive ailment. Equally importantly, Garrod noticed that the more protein his alkaptonuric patients ate, the blacker their urine became, which suggested one thing to him: his patients must be unable to carry out the final step in breaking down an essential component of the protein in their diet. It was an accumulation of protein breakdown products that was darkening his patients' urine and was causing their arthritis and nose and ear discolouration. This explanation implied that victims must have inherited a lack of an enzyme which normally carries out this final breakdown, a deficiency that was giving rise to the 'inborn error of metabolism' of alkaptonuria. It was an inspired piece of detective work that involved chemical studies, the presumption of an enzyme defect, and a clear understanding of patterns of inheritance. Indeed, Garrod's hypothesis was so advanced that it was not fully confirmed for another forty years. Once again, a brilliant insight was followed by decades of neglect and lack of peer recognition – just as it was for Mendel, Herrick and other pioneers of genetics.

The discovery of alkaptonuria was important because for the first time an inherited defect could be directly related to a change in the body's chemistry, in this case a malfunctioning enzyme, so providing a clue in the search for the basic defects of other inherited ailments, such as sickle cell anaemia. In this case, researchers wanted to know: is there a parallel protein whose malfunction explains the sickling of red blood cells?

As we saw in the previous chapter, blood would be virtually colourless were it not for its oxygen-carrying red cells. If you isolate these cells by letting them settle to the bottom of a test tube, and then add plain tap water, they will burst open, releasing their major protein constituent: haemoglobin. (Living cells always burst open in pure water because of salt imbalances set up between the liquids outside and inside their membranes.) And if you bubble oxygen through a blood sample, it will turn from a dull almost blueish red to a very bright red – because haemoglobin carries oxygen round the blood and what you are doing in that simple experiment is to mimic what happens when blood is pumped by the heart through the lungs to pick up oxygen, and take it on to the tissues of the body. Oxygen-rich haemoglobin is bright red but turns more dull and blueish in colour after it has delivered its oxygen. And since it was oxygen deficiency that seemed to give sickle-trait red cells their shape, it seemed obvious that haemoglobin should be the candidate for the malfunctioning protein in sickle cell anaemia.

Well, with hindsight, that may seem to be the case. It nevertheless took considerable efforts by researchers before two scientists – William Castle, professor of medicine at Harvard Medical School, and the Nobel Prize winning chemist, Linus Pauling – confirmed the hypothesis. Castle's early experiments had suggested that a protein abnormality caused red blood cells to sickle and Pauling interpreted this result as implicating haemoglobin. So he directed three of his young associates to discover if sickle cell anaemia patients' haemoglobin was different from normal haemoglobin. They eventually succeeded by using electrophoresis, a technique in which proteins of slightly differing electrical properties can be separated when an electric current is applied to them.

Pauling and his team found there were two forms of haemoglobin – normal and sickled. Individuals with full sickle cell anaemia had only abnormal haemoglobin; carriers had a mixture of normal and abnormal haemoglobins, while normal individuals had, of course, only normal haemoglobin. Sickle-cell-trait carriers presumably then had enough normal haemoglobin to prevent them from getting any signs of anaemia, but enough of the abnormal haemoglobin to show sickled red cells when their blood was deprived of oxygen on a microscope slide. It was an important discovery which for the first time directly demonstrated a protein abnormality that was precisely associated with a mutation in a gene.

But how could genes and these multifarious, complex proteins be connected? How did genes manifest themselves? Indeed, was it possible that genes themselves were proteins? In the inter-war years, this last idea seemed quite natural. After all, chromosomes were known to contain proteins, though they were also known to have other constituents.

The answer was provided by experiments in which proteins, heated in either a strong acidic or strong alkaline solution, were found to break down into twenty different, much smaller, molecules called amino acids. These amino acids are the building blocks of proteins. They join together in long strings called polypeptides to form proteins which can vary enormously in size. Some contain a few dozen amino acids. Some have thousands. And since there are twenty different amino acids, there must be many, many possible protein sequences. After all, many different words can be constructed from a twenty-six-letter alphabet. A good English dictionary typically contains well over thirty thousand entries, and very few of these are more than ten letters long, never mind a couple of thousand. So we can see that the Book of Man is a big read by any standards!

But did a protein have a well defined structure, or was it just a jumble of different amino acid combinations, scientists wondered. The answer was not resolved until the great British biochemist, Fred Sanger, painstakingly

sequenced the first protein: insulin. Sanger spent ten years carefully identifying small fragments of the insulin molecule and working out their
constitution until he had pieced it all together. Sanger had chosen insulin
because it was then the only protein that he could buy in relatively pure
quantities. 'That was lucky because insulin has since turned out to be one of
the smallest proteins and was therefore one of the easiest to sequence,' he
recalls. Nevertheless, it took a decade to break apart the insulin molecule,
then painstakingly separate all its closely related chemical components
before rearranging them in their correct sequence. 'It was like trying to find
out what a car was made of by taking a sledgehammer to it,' says Sanger.
'Then, when you've bashed it into little pieces you try to put these together
again.'

In the end, using a newly developed technique called partition chro-
matography, Sanger showed that the insulin molecule consisted of two
chains of amino acids. One contained a sequence of thirty amino acids and
the other a different sequence of twenty-one amino acids. The discovery was
striking because it showed that a protein is a unique sequence of amino acids
which give it a distinct three-dimensional structure upon which its properties depend. In addition, once the structures of proteins – the machinery of
living matter, as Sanger calls them – became known, it was then possible to
ask: how are they made? Answering this was to lead researchers into one of
the greatest quests in scientific history – the unravelling of the stuff of genes,
the first step in the setting up of the Human Genome Project itself.

What Sanger showed was that we can look at proteins as biological words
made up from a twenty-letter alphabet, with each word having a different
meaning or function. And when there are misprints in a word then its meaning may be changed or lost. In the case of haemoglobin, consisting as it does
of a total of three hundred amino acids, there is only one letter difference, a
single misprint, between the normal and sickle versions, yet the outcome, as
we have seen, can be deadly, triggering a cascade of effects through the body
to produce the symptoms of sickle cell anaemia. The genetic typing error
occurs in the sixth amino acid in one of haemoglobin's two chains, the beta.
Glutamic acid is replaced by valine. This alters the chemical bonds within
the protein, which becomes curved or hook shaped, and so the red blood
cells tend to catch each other, eliminating red blood cells' ability to squeeze
through blood vessels. The result is anaemia.

It was this complexity of proteins, and the amount of information they
contain, that led some scientists to think that genes might be made of proteins. But the essence of genes and of life itself is reproduction and duality,
as we saw in the last chapter. Genes have to make precise copies of themselves each time a cell divides, and there was nothing in the structure of

proteins to suggest they were capable of this process. So, to find out what genes are made of, scientists turned to an avenue of research that had been started in the mid-nineteenth century.

In those days it was known that an infected, festering wound accumulated pus. Much of this pus, we now realize, is made up of white cells in the blood which gather there to counteract infection. Pus-soaked bandages therefore made an excellent, if unappetizing, source of white cells; one that a young Swiss chemist, Johann Miescher, decided to exploit to uncover the chemical make-up of cell nuclei. At that time, in 1868, the nucleus was merely viewed as a scarcely visible blob in the centre of a cell. The nucleus's role as the repository of chromosomes and hereditary information had still to be defined.

When Miescher investigated the nuclei obtained from pus-soaked surgical bandages, he found a new compound that contained large amounts of phosphorus and which he called 'nuclein', because he had found it in cell nuclei. The substance was so unusual that Miescher's teacher, Hopper Seyler, then editor of the journal in which Miescher's paper was to be published, held back publication for two years while he repeated some of the young researcher's 'improbable' observations. In particular, Seyler was surprised at the presence of phosphorus at the heart of a living cell. The element, he felt, had no right to be there. A short sojourn at the laboratory bench disabused him of this notion, however.

After he had been vindicated, Miescher and his colleagues began to work out the strange new substance's basic chemistry and found it was acidic in nature. So it was rechristened nucleic acid. Then it was discovered that there are two varieties of nucleic acid, one called deoxyribonucleic acid, DNA, and another called ribonucleic acid or RNA. The former, DNA, was shown to contain four essential chemical constituents: adenine (A), cytosine (C), guanine (G) and thymine (T), a four-letter set that we know today is the alphabet of the gene. In the case of RNA, thymine (T) is replaced by uracil (U). At first, it was thought that DNA was found in animals and RNA in plants, but it was soon realized that DNA is present at the heart of the cells of all living organisms.

This chemistry was being worked out just as Mendel and his laws were being rediscovered and the other fundamental breakthroughs of genetic science, discussed in Chapter Two, were being made. There seemed little connection between the chemistry of DNA and the fundamental laws of inheritance, however, though Miescher must have had some inkling when he wrote, in 1892, in a letter to an uncle, that the large molecule he was working on might convey a hereditary message 'just as the words and concepts of all languages can find expression in twenty-four to thirty letters of the alphabet'.

In general, however, researchers were uninterested in DNA – which may seem odd in retrospect. After all, by then the stuff had been shown to be present in the nucleus and in the chromosome, which was known to be the seat of genetic information. Surely the signals were strong? The trouble was that several respected chemists, such as Phoebus Levene of the Rockefeller Institute of Medical Research, were opposed to the idea that DNA was important. Levene argued that DNA was just a monotonous molecule containing strings of arbitrary repeats of those four basic letters – G, A, C and T. Not for the first or last time did the views of a conservative authority inhibit scientific progress. Indeed, as in other fields, scientific ideas and experiments tend to follow prevailing fashions, set by a consensus that is strongly influenced by distinguished figures of authority. Critical new discoveries often come from those, mostly younger, scientists who are prepared to break the established mould. As we have already seen, the greatest revelations in genetics combined ideas and methods from quite different scientific areas, even different disciplines – witness the work of James Herrick and Victor Emmel. And it was to be the combination of rigorous chemical research with biological insight that helped scientists make their next step in showing that DNA was the carrier of genetic information. Later, the entry of physicists into the field, bearing their specialist knowledge and techniques, propelled science even further down this road, revealing how DNA actually carries its message.

But before we run away with ourselves, let us look at how DNA was shown to be the 'stuff of genes'. And once again we find clues coming from unexpected directions; not from humans, not even from animals, but from studies on bacteria. Just as the horse threadworm had been instrumental in the discovery of chromosomes, so another lowly form of life played its strange part in the translation of the language of life.

At this time, in the early 1920s, before antibiotics had been discovered, pneumonia – caused by the bacterium pneumococcus – was a major killer, and was naturally the focus of considerable research. Scientists were particularly interested in trying to immunise people with a weakened or 'attenuated' strain of bacteria that would not cause the disease but would protect against future infection. It was in the midst of such research that Frederick Griffiths, working at the pathology laboratory of the Ministry of Health in London, discovered two forms of the bacteria, which he simply labelled smooth and rough according to their appearance. The rough form was a weakened one. Injected into mice, it did not kill them. On the other hand, the smooth type was virulent and deadly, although when killed by heat and then inoculated, it too was rendered harmless.

But when Griffiths blended one strain of his rough, innocuous bacteria with a strain of the virulent smooth bacteria that had been heat-killed, he made an astonishing discovery. The mixture killed his laboratory mice. There it was – a concoction of two innocuous strains, one living but ineffective, the other heat-killed, suddenly gaining the power to slay. It was as if two simple culinary ingredients when mixed in a cake acquired lethal powers. Somehow, the heat-killed, smooth bacteria had transferred their virulence to the living, rough strain, turning them into pneumococcal killers. Moreover, the change was permanent and implied that a distinct genetic modification had taken place. What was causing this startling transformation?

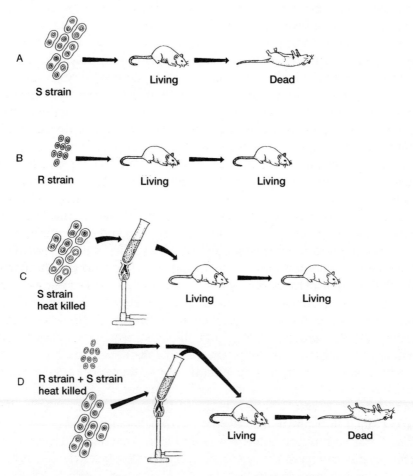

Griffith's classic experiment: A mouse injected with smooth, virulent bacteria (A) is killed. A mouse injected with rough, innocuous bacteria (B) survives. A mouse injected with smooth, heat-treated bacteria (C) also survives. But a mouse injected with a mixture (D) of bacteria types B and C is killed.

The answer was provided, eventually, by Oswald Avery at the Rockefeller Institute in New York. When he read Griffiths' account of his bacterial transformations, he was sceptical. The experiments appeared to violate the notion of stability of species. But after he repeated some of Griffiths' experiments, Avery realized the phenomenon was real and he decided to discover the chemical identity of the 'transforming principle'. It took Avery and his colleagues, Colin McLeod and Maclyn McCarty, using a test-tube version of Griffiths' experiment, fourteen years to compile convincing evidence. The answer, of course, was that DNA was carrying out the transmission.

Avery had long toyed with the idea that DNA was responsible, but the eventual proof depended on exhaustive purification and characterization of bacterial extracts. To persuade sceptics, he had to show that he had removed every trace of protein or any other contaminant which might be responsible for the transformation. In a letter to his medical bacteriologist brother Roy in 1943, Avery described how on 'drop wise addition of absolute ethyl alcohol' you can precipitate out 'a fibrous substance, which on stirring the mixture wraps itself about the glass rod like a thread on a spool' leaving behind other impurities. We know that those diaphanous strands are the threads of life itself: DNA.

Avery clearly realized the momentous nature of his work; as far as his bacteria were concerned, their genes appeared to be made of DNA. However, his final, published paper, as is so often the case in the stylized, hesitating language of the scientist, ended much more coyly. In that he stated that perhaps the biological activity was 'not an inherent property of the nucleic acid but is due to minute amounts of some other substance absorbed into it, or so intimately associated with it as to escape detection.' However, Avery did add: 'If the results of the present study . . . are confirmed, then nucleic acids must be regarded as possessing biological specificity, the chemical basis of which is as yet undetermined.' And therein lay the next challenge – to establish the chemical structure of DNA and, through that knowledge, gain clues as to how it operated as the basic material that makes up a gene.

By the time Avery had completed his work, the Nobel Prize winning British organic chemist, Alexander Todd, had established how its chemical components were linked. He showed there was a backbone made from alternating sugar and phosphate groups, with each phosphate, a group of phosphorus and oxygen atoms, linked to the next through a sugar. (There are many different types of sugars, of which sucrose, which we put in our tea or coffee, is only one type. In the case of DNA, the sugar is a type known as deoxyribose.) In addition, one of the four bases – A, T, G or C – was attached to each sugar on the opposite side to the phosphate link. But this simple description of a string of bases held together by sugars and phosphates merely revealed a rough pattern. It left completely unsolved the question of what was the underlying three-dimensional structure. How many chains were there? Did they associate with each other, and if so how? And most importantly, how did the structure of DNA relate to its behaviour during cell division, when it passed on the genetic blueprint of an entire creature? The answers were given by James Dewey Watson and Francis Crick in a paper of less than one thousand words in the English scientific journal *Nature*, published on 25 April 1953.

Francis Crick had graduated as a physicist, but after working for the British Admiralty during World War II had decided he wanted to apply the rigour of physical analysis to the study of living material. He ended up working at the Cavendish Laboratory in Cambridge where scientists were trying to use X-rays to study the structure of proteins. As a newcomer to biology Crick appreciated that the key lay with genetics. It was essential to relate the chemical structure of the gene to the structure of a protein, based on its sequence of amino acid letters, he realized.

His partner, James Watson, a man whose name was to be linked with Crick's as inextricably as Cain's to Abel's, arrived at the Cavendish with a very different pedigree. He had entered Chicago University to study biology in 1943 as a precocious fifteen-year-old, before moving to Indiana University to do his doctoral work on the genetic analysis of bacterial viruses. Then he moved to Europe, and after an unsatisfactory period studying the biochemistry of nucleic acids in Copenhagen, he decided that unravelling the structure of DNA would be his goal. And so it was that he joined Francis Crick in Cambridge, with the aim of using X-ray techniques as his first route of attack. As Watson states in his racy account of their endeavours, *The Double Helix*, 'finding someone . . . who knew that DNA was more important than proteins was real luck.' But there was more to the relationship than mere shared interest, as Crick recalls. 'Jim and I hit it off immediately, partly because our interests were astonishingly similar and partly, I suspect, because a certain youthful arrogance, a ruthlessness, and an impatience with sloppy thinking came naturally to both of us.'

The pair were also fortunate in their choice of setting. At the Cavendish, Crick and Watson worked under Lawrence Bragg, who had shown that when X-rays were shone on to a pure crystalline chemical, they could be used to reveal its three-dimensional structure. For that achievement the twenty-five-year-old Bragg and his father, William, had shared the Nobel Prize in physics. Then in 1938, Bragg junior succeeded Lord Rutherford, the atom splitter, as the head of the Cavendish Laboratory, where he inherited an array of talent of startling intensity, including the charismatic John Desmond Bernal, a founding father of X-ray crystallography, and Max Perutz, who

The discoverers of the structure of DNA: James Watson (left) and Francis Crick (right).

shared the Nobel Prize with John Kendrew in 1962 for establishing the structure of a protein molecule using the X-ray technique. From these small beginnings – in which techniques newly honed by the physicist were turned on the study of living processes – the Cavendish Laboratory for Molecular Biology developed. Since then the laboratory has been the home of a breathtaking number of fundamental discoveries in biological and medical research, to the extent that its researchers had amassed a total of eight Nobel Prizes by 1984.

But none of the laboratory's scientific exploits will ever rival those of Jim Watson and Francis Crick who solved the structure of DNA, a tale that has since been told in books, films and television programmes. The story, characterized by scientific opportunism, brashness and brilliant deduction, catches the imagination not only because it makes a good yarn, but because the result was the most important milestone in our understanding of the living world since Darwin devised his theory of natural selection and Mendel discovered his laws of inheritance. Yet Crick and Watson never took an X-ray picture of DNA, or carried out a significant experiment. They merely looked at, and thought about, the work of others, such as Maurice Wilkins and Rosalind Franklin.

Maurice Wilkins, like Francis Crick, trained as a physicist and worked on the Manhattan project during the war, an experience that left him with an abiding revulsion to the A bomb. So he moved to the study of living things, at King's College, London, and after learning X-ray crystallography turned to the investigation of the structure of DNA. He was joined by Rosalind Franklin, a chemist who brought extraordinary skills in the preparation of material and the implementation of experiments. She had already taken the best X-ray pictures of DNA that anyone had yet obtained. Unfortunately Wilkins and Franklin did not get on, and although Wilkins was the senior partner in the collaboration, Franklin clearly felt that the DNA project had been assigned to her. The situation was not helped by the fact that women were not made particularly welcome in academic circles in those days, and Franklin was clearly highly intelligent, determined and competitive. It was she who provided the pictures from which Crick and Watson deduced their famous double helical structure for DNA. Sadly Rosalind Franklin died of leukaemia in 1958, when she was only thirty-seven, and so was not alive when Crick, Watson and Wilkins shared the Nobel Prize in 1962 for their discovery of the structure of DNA. (The Nobel committee is restricted to awarding its prize in any given subject to no more than three people at a time. One wonders what they would have done had Rosalind Franklin been alive.)

At the time, helices were all the fashion, since Pauling had shown that many proteins had spiral structures. Crick had been working in this area,

Rosalind Franklin.

and so knew precisely what sort of X-ray photo would be produced by large helical molecules. Recognizing just such a component in the structure of DNA was the first critical clue – provided by a clandestine peek at one of Rosalind Franklin's best photographs.

Such behaviour was typical of Crick and Watson. For them, the prize – to be the first to unravel the structure of DNA – was simply 'up for grabs', in Watson's words. The niceties of academic procedure were of no concern to them. And although there was a fair amount of communication between the groups in Cambridge and London, there was a definite difference of philosophy between them. Watson and Crick were the theoreticians and model builders, while Wilkins and Franklin emphasized the primacy of experimentation and the analysis of data over mere 'speculation' – a classic division of talents to be found in all other sciences.

It was clear to Crick and Watson that the structure must contain helices. But how many? Either two or three, they guessed. Looking back it seems surprising that the latter notion of a triple helix was ever seriously entertained, given the duality of the reproductive process. Once it had been accepted that the structure consisted of two intertwined helices, the key question was: how do the bases adenine, cytosine, guanine and thymine, our four-letter A T G C alphabet which stretches along DNA's two helical strands, interrelate with each other? An obvious possibility was that like paired with like, so that an A went with an A, a G with a G and so on. But this was soon proved to be chemically untenable. The shapes of the bases just would not let them fit together.

The puzzle was solved through a simple observation of the relative amounts of A, T, G and C in DNA. Scientists had already discovered that the amount of A was always equal to the amount of T, and the amount of G equal to the amount of C. This rule gave Watson the clue as to how the two helices of their predicted structure fitted together. He reasoned that opposite an A on one strand there was always a T on the other, while opposite a G on one strand there was always a C on the other. This pairing rule fitted precisely the known shapes of the four bases which fell into two pairs, the purines – A and G – and the pyrimidines – C and T. The pairing rule always placed a purine opposite a pyrimidine. In this way, the distance between the backbones of the two helices is kept constant, so maintaining a regular structure that could form a crystal.

And what a crystal it was. This, as Watson described it, was 'the Rosetta Stone for unravelling the true secret of life'. What the two researchers had found was that DNA is shaped in a double helix, a miniature spiral staircase, with the long chains of sugars and phosphates acting as the banisters and the linked pairs of adenine–thymine and cytosine–guanine groups forming the

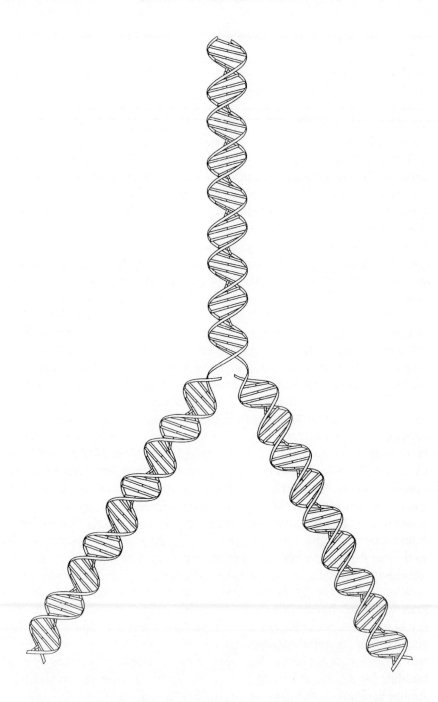

How life comes from life. The division of DNA into two separate strands on to which grow two new strands is the 'copying mechanism for genetic material'.

steps. This is the 'golden molecule' that controls how we will grow and develop.

One of the first groups to whom Crick and Watson showed a model of the double helix structure of DNA was Wilkins and Franklin. They had, after all, been resistant to Watson and Crick's speculations. Yet they were instantly struck by its beautiful simplicity. As Watson put it, even Franklin 'accepted the fact that the structure was too pretty not to be true'.

The particular grace of the double helix lay with its structure which suggested that it could duplicate itself quite easily. If its two strands separated, then new strands could build along the old following the pairing rule A–T and G–C, to create two new identical daughter helices. For instance, a thymine base would attach itself to an adenine, as that is the only chemical base to which it can attach itself. In other words, the double helix becomes two single strands during cell division, each of which then grows a second, complementary strand by a process that we shall discuss in the succeeding pages.

Clearly, this four-letter alphabet of A, C, G and T forms life's mother tongue, the true language of reproduction. As Crick wrote to his son Michael: 'We think we have found the basic mechanism by which life comes from life.' However, when the two researchers got round to writing their paper for *Nature* they were somewhat more circumspect. They merely stated that 'it has not escaped our notice that the specific pairing we have postulated immediately suggests a possible copying mechanism for the genetic material.' This copying mechanism operates by dividing DNA into its two strands. On each of these grows a second strand. The exact means by which the component As, Cs, Gs and Ts were placed upon their complementary bases was not obvious, however, and in the end it took Arthur Kornberg's discovery of an enzyme in bacteria which could duplicate DNA to provide the answer.

Kornberg was a passionate advocate of enzymes. He once admitted that he abandoned his animal nutrition laboratory 'when I realised enzymes are the vital force in biology, the sites of vitamin action, and the means for a better understanding of life as chemistry'. Fortunately the love affair was a reciprocal one, in a sense, for Kornberg's 'vital forces' were to lead him to a Nobel Prize which he shared with Spanish scientist Severo Ochoa for their discovery of nucleic-acid-duplicating enzymes. Kornberg was able to purify an enzyme from bacterial extracts that, when given some DNA as a template and the relevant building blocks, made new molecules from the added building blocks which matched precisely the DNA template. Ochoa, using similar techniques, did the same with RNA.

Kornberg and Ochoa's duplicating enzymes helped solve another fundamental problem – the nature of the genetic code. Somehow the sequences of the four bases A, C, G and T that stretch along DNA molecules must code for the sequences of the twenty different types of amino acids that make up proteins. This is the only way to envisage a direct relationship between a gene and a protein which the pioneering research outlined earlier in this chapter had shown must exist. The challenge was to decipher the code that related these two sets of sequences.

It was obvious that a single base pair was inadequate to delineate an amino acid. There are only four of the former – A, C, G and T – and a total of twenty different varieties of the latter. Similarly, a pair of base pairs is inadequate to specify an amino acid, for there are only sixteen varieties (4 × 4) of these. The answer, it then transpired, was a triplet combination of base pairs. As we can see there are sixty-four possible combinations (4 × 4 × 4) of these, enough to account for all twenty amino acids and perhaps some punctuation marks, with plenty of spare capacity left over. (In fact, some amino acids are coded by more than one triplet combination, or codon as they are known.) This idea of a triplet code was subsequently confirmed by a series of elegant genetic experiments carried out by Francis Crick and Sidney Brenner, although a great deal of hard, sophisticated, experimental work was still needed finally to establish the different amino acids that are coded by different triplets of base pairs. (As an example, the triplet adenine–adenine–guanine codes for the amino acid lysine.)

One by one pieces were being put into position in our biological jigsaw, though one very basic question still had to be answered: how exactly did a particular DNA section, i.e. a gene, get transformed within a cell into its corresponding protein? There seemed to be no way in which DNA could itself act directly as a template for assembling the amino acids in a protein. There was simply no physical correspondence between the DNA and the amino acid. In addition, experiments showed that it was the other form of nucleic acid, RNA, which was mainly found within a cell's cytoplasm and which seemed to be directly involved in protein synthesis.

The solution arrived when scientists discovered that a small fraction of this RNA seemed to be broken down and degraded almost as soon as it was made, but which had the property that its composition of bases, with U replacing T, was just the same as that in DNA. In other words, here was highly suggestive evidence that an RNA copy was being made of a DNA sequence, indicating that there was a definite pathway in which DNA made an RNA copy of itself and that this replica in turn acted as the template for the protein. The idea was that special DNA-like RNA, called messenger RNA, was actually the molecule which controlled the construction of a protein's amino acid sequence.

It was left to Francis Crick to put forward the idea of an 'adaptor' mole-

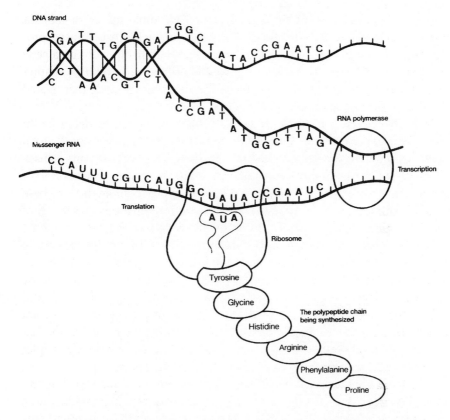

How the lexicon of life is read. The manufacture of a protein begins when a segment of DNA unwinds and one of the strands provides the code for the synthesis of messenger RNA, a process that is known as transcription. This messenger RNA then undergoes translation into an amino acid on a ribosome. Each triplet of RNA codes for a specific amino acid, and these amino acids combine to form a polypeptide chain or a protein.

cule which transforms this template into a sequence of amino acids. These, we now know, carry a particular amino acid and latch on to RNA sequences. There is an adaptor molecule for every codon, and these swim around the cytoplasm waiting for the right RNA sequence to heave into view. When it does, the adaptor molecule latches on to the RNA and delivers its amino acid in response to this contact. A ribosome (which is also made of RNA) holds the adaptor molecule in place and so acts as the engine that drives the translation of DNA into proteins via messenger RNA. This idea was eventually proved to be correct by Marshall Nirenberg, a biochemist, who used radioactively-labelled amino acids and synthetic RNAs to make proteins with cell extracts in the test tube, so proving that Crick's original, inspired idea was correct.

An example of how this process works is provided by the amino acid phenylalanine. This is coded for by the triplet UUU. The adaptor molecule for phenylalanine has the sequence AAA at one end, the complementary stretch of RNA to the AAA codon. When it latches on to it, the adaptor molecule delivers the amino acid phenylalanine.

The whole system is simple and elegant. The double helical DNA molecule carries genetic information in its sequences of A, T, G and C letters. Duplication, the essence of life, happens when the two helical strands separate and new strands grow along the old sequences, producing two identical copies from one DNA double helix. Then when the DNA is primed to make a protein, a messenger RNA copy is made, and this forms a template using small RNA adaptor molecules that read off codons of three letters to deliver the appropriate amino acid. These amino acids then combine to form a protein. This process explains directly the relationship between genes and proteins. It also explains why things can go wrong. Mutant abnormal pro-

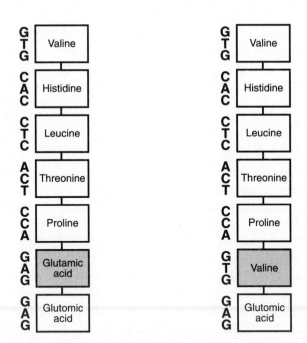

The sickle cell mutation. One amino acid out of a total of 287 differs in sickle cell haemoglobin compared with normal haemoglobin. In the latter, the triplet GAG (guanine, adenine, guanine) codes for the amino acid glutamic acid (left-hand diagram). In sickle cell haemogobin, the corresponding three-base sequence is GTG (guanine, thymine, guanine) which codes for the amino acid valine. This substitution occurs sixth in line from one of the ends of the beta haemoglobin chain.

teins will be produced when there is an error in a DNA sequence. A perfect example is provided by the sickle cell trait and the anaemia that is caused by abnormal haemoglobin.

Direct analysis of the abnormal sickle cell haemoglobin has shown that it differs from normal haemoglobin in just one of its 287 amino acids. Normal haemoglobin actually consists of two different polypeptide chains of amino acids, called an alpha and beta chain. These two chains are made from two different genes. The alpha chain is 141 amino acids long. The beta chain has 146 amino acids. And the only difference between sickle and normal haemoglobin is that the former has the amino acid valine, instead of glutamic acid, in position 6 of its beta chain. And that potentially lethal difference is due to just one base-pair difference at the DNA level, namely a switch from an A to a T which changes the triplet GAG (which codes for glutamic acid) into GTG (which codes for valine).

And that is it. Just one spelling error along the three thousand million DNA letters that make up a person's genome ('genome' is the name we give to a cell's, or a person's, total compliment of chromosomes) is enough to give rise to abnormal haemoglobin that can confer malarial resistance when combined with normal haemoglobin, but which gives rise to severe anaemia on its own.

The delicate molecular detective work that has revealed this picture has since been used to uncover the roots of dozens of other inherited diseases, ailments that once seemed intractable and almost inexplicable. Thanks to the work of teams of scientists who have followed in the wake of Mendel, Garrod and the rest, we can now explain these illnesses in terms of alterations that have occurred in a person's DNA sequence, mutations that produce deviations in the proteins they control, so triggering all the symptoms of an inherited condition. In the next chapter we shall trace how modern biological engineers have used this newly acquired ability to translate what was the impenetrable language of life in order to create technologies that allow them to tinker with the once inviolate nature of living organisms. New medicines, screening techniques and breakthroughs in understanding diseases have followed with startling speed.

4

Lords of the Genome

Walter Gilbert and Fred Sanger stand at the very opposite ends of a scientific spectrum. On the one hand Gilbert is a power broker, a man of grandiose aims, ranging from undertaking fundamental genetic research to attempting to exploit the entire human genome commercially. His surroundings are correspondingly august: large offices, leather chairs, chrome tables and other appurtenances so beloved of modern executives. By contrast, Sanger, the Nobel Prize winner who first revealed the structure of a protein, is a quintessential stereotype of the dedicated researcher, a man who has eschewed glamour for modest obscurity, absorbing himself in his research.

Yet for all their differences in character, these men played strikingly similar roles in the decoding of the Book of Man, following paths that led them to the same goal, the 'holy grail' of genetics, as Gilbert called their unravelling of the string of bases that make up a gene. In doing so, Gilbert and Sanger helped found a dynasty, the 'Lords of the Genome', who would transform genetics from a science of descriptive analysis into a powerful, manipulative technology. They became the first scientists to create new lifeforms that had never been anticipated by natural selection. They began to play God, in a sense. This astonishing tale begins with those genetic pathfinders, Gilbert and Sanger; particularly with Fred Sanger.

Sanger is, above all, a scientist with an extraordinary gift for developing methodologies, for perfecting techniques that can crack seemingly intractable biological problems. Given a choice of thinking, talking or doing, 'I much prefer the last and am probably best at it,' he says. 'I am all right at the thinking, but not much good at the talking.' Not for Sanger then the heady thrill of theorizing, or the frisson of speculation with colleagues. Until his retirement in 1983, he was happiest when actually carrying out an experiment.

Frederick Sanger, British biochemist and double Nobel laureate.

Walter Gilbert, Harvard University Nobel laureate scientist.

'My talents have always been in developing new methods that are essential for tackling biochemical problems,' he says. And as we have already seen, he has been unquestionably effective at that task. His splintering of insulin into its amino acid components which he then reassembled to reveal the protein's structure earned him a Nobel Prize, a mighty achievement that would satisfy any scientist. But Sanger was not finished, and his next moves reveal just how gifted and original a thinker he was. Let us examine them in some detail, for they show that the process of research, as well as being painstaking and methodical, can also be breathtakingly imaginative and intellectually creative, a fact of which the public is often sadly unaware.

Having worked out the structure of a gene product, in other words a protein, his next labour – the unravelling of a gene itself – would seem to be a fairly logical step, at least in retrospect. Sanger was not so sure. Proteins still had a lure for him, and he continued to tinker with them. Then, in 1962, he went to work at the new laboratory of molecular biology that had just been built at the edge of Cambridge and there he joined Perutz, Crick, Brenner and several other noted 'gene hunters'. The move was decisive. 'With people like Francis Crick around, it was difficult to ignore nucleic acids or to fail to realize the importance of sequencing,' Sanger recalls.

In the last chapter we saw how Crick and Watson had shown that those adenine (A), cytosine (C), guanine (G) and thymine (T) bases were the building blocks of genes, and that triplets of them coded for specific amino acids. That is the alphabet of the genome. But what order do these letters take when they form genes, the words that make up the Book of Man, scientists wondered. To answer that, researchers would have to develop a way of rapidly reading the order of As, Cs, Gs and Ts as they appeared along a piece of DNA. Until they could do this, they could not begin to unravel the byzantine complexity of the genome and reveal how it controlled the physiology of a human being.

The task was not simple. In particular, there was the question of interpretation of results. Because DNA has only four essential chemical constituents, only very large sections would provide distinctive enough regions to allow overlaps to be made and full sequences worked out. However, by using the tried-and-tested approach that Sanger had exploited in his insulin work, scientists at various centres began to sequence different pieces of nucleotides (stretches of genetic material). But progress was painfully slow. 'The main problem with DNA was its very large size. The smallest pure DNAs that were available were genomes of single-stranded bacteriophages [small bacterial viruses, or viruses which attack bacteria] of 5000 nucleotides,' says Sanger. And given that researchers were progressing at rates of a few dozen A, C, G and T nucleotides a year, Sanger quickly realized that his old divide-and-recombine method that had cracked the structure of the insulin protein was not going to be good enough for DNA.

So he simply turned the problem on its head. If breaking down DNA into sub-units, then piecing them together until a full sequence had been created, would not work, he would start from scratch. He would build up his DNA virtually from zero. And the tool that he selected for his task was the enzyme called DNA polymerase.

DNA polymerase is one of the DNA-copying enzymes that we met in the previous chapter, and it is responsible for stitching together those lined-up A, C, G, T building blocks into a second strand of DNA against the pre-existing template. There it helps build a second strand against the template. Sanger decided that if he could interrupt the polymerase enzyme as it carried out this work, he might be able to use the information to decipher the sequence of that strand.

So he mixed his building blocks – adenine, cytosine, guanine and thymine – with single-stranded DNA and polymerase, and separated this concoction into four flasks. But crucially, in each case, Sanger placed insufficient amounts of A, or C, or G, or T into each flask in turn, which meant that in each of the four flasks his second strand stopped growing at various points because there was not enough of one or other of the bases. Say we consider the reaction with insufficient adenine. At any point where an A should be put into place by the polymerase, the growing strand might stop for want of an adenine molecule. Now all the second strands had started growing from the same starting point and some would stop early and some would stop late, depending – as chance dictated – when they had run out of adenine. In theory, the end result would be the creation of an entire library of DNA segments all ending at an A. In other words, a set of fledgling DNA second strands, all ending at an A, should be created, and by grading their lengths, it would be possible to work out where each A occurred in relation to the starting point of the polymerase reaction. Similar sets for C, and G, and T could also be created and the fragments compared in size.

It was a neat idea. Its only drawback was that it did not work very well. The separation of the different fragments was poor, and the rate at which DNA sequences could be read off was painfully slow. The process seemed sound, but its implementation proved awkward. It was then Sanger had the idea that was to be 'the climax of my research career', as he subsequently put it. Instead of starving his reaction of As, Cs, Gs and Ts, he would add chemicals called dideoxy, or chain, terminators. These were altered versions of adenine, cytosine, guanine and thymine and they had one special property. When one of these biochemical imposters was inserted by polymerase, instead of a normal base, it jammed the copying process at that point.

The effect is random, and is like creating an assortment of DNA drop-sticks, each a different length, but always ending in the same letter, depending on which base and corresponding terminator that you use. The effect was the same as that produced by Sanger's previous attempt at sequencing, but turned out to be a lot more effective. Then by sifting through the different dropsticks and comparing their lengths, you can work out where each A must lie on a strand and similarly for the other bases.

Let us look at one small scrap of DNA, as an example; a sequence that contains only ten bases and which runs, from beginning to end, in the following way: AGCTATGGAC. If we try to grow this on a template but with the added condition that we add A terminators, we will find that growth will stop at positions 1 or 5 or 9 depending where the adenine terminator randomly attaches itself. Similarly for cytosine, it will attach positions at 3 and 10; guanine at 2, 7 and 8; and for thymine at 4 and 6. In other words, when we put adenine terminators in, we get DNA sequences that are 1, 5 and 9 units in length; with cytosine, we get lengths of 3 and 10 units; with guanine, we get 2, 7 and 8; and thymine, we get 4 and 6. All we then have to do is sort out these different DNA dropsticks and grade them for length, and then we will have obtained the exact sequence of that scrap of DNA.

This last task, analysing the DNA dropsticks for length, is the easy part. The DNA mixtures are put into a slot of a polyacrylamide gel, which is a bit like a fruit jelly, in a tray with slots at one end, and an electric field is applied to the opposite ends of the gel. Negatively charged DNA is attracted towards the positive pole, with the smaller fragments being pulled through faster than the larger ones. After a while, the pieces spread out and can be graded by size. The gel is then treated with an appropriate stain to highlight the DNA fragments, or is labelled with a radioactive probe so that a permanent image can be created on a film. By placing the dropsticks in four different lanes, one for each A, C, G or T terminator, it then becomes a very simple matter for the experimenter to read off the DNA sequence from the bottom of the gel to the top.

In essence, the procedure is absurdly simple, once you think about it – the hallmark of a truly original idea. Using his terminators, Sanger and his team were able to read off chunks of a bacteriophage, phi-X-174, at a rate of 300 bases at a time. These genetic nuggets were then overlapped and aligned according to the old divide-and-recombine methods that Sanger had used on the insulin protein. The phi-X-174 sequence was published in *Nature* on 24 February 1977, and took the world of molecular biology by storm. Sanger had sequenced the first genome (admittedly a very simple one) and in doing so paved the way for the mighty assault that molecular biology has subsequently made on biological and medical problems over the past two decades. For his pains, Sanger was awarded a second Nobel Prize, in 1980, placing him in a very small and select band of individuals. Only Marie Curie and the physicist John Bardeen (inventor of the transistor and the theory of superconductivity) have been similarly doubly rewarded for achievements in science (though we should not forget Linus Pauling, who was given the Nobel Prize for chemistry, and later the Nobel Peace Prize). Yet Sanger remains ridiculously modest about his exploits. 'I am not academically brilliant,' he claimed in a rare autobiographical article published in 1988. 'I never won scholarships and would probably not have been able to attend Cambridge University if my parents had not been fairly rich.' Perhaps, but he remains one of the greatest experimentalists of the twentieth century just the same.

This brings us to Walter Gilbert, the man who stood beside Frederick Sanger in December 1980, at the Nobel Prize ceremony in Stockholm. Gilbert was there for exactly the same reason as Sanger. He too had led the way to the unravelling of the gene, though by a very different method.

Gilbert was originally a theoretical physicist, but switched to biology in 1960 at the age of twenty-eight. Within a few years his distinguished work had earned him a chair, sponsored by the American Cancer Society, at Harvard University. After this, Gilbert's first major contribution to molecular biology was the discovery of a protein known as a repressor, a biological entity whose existence had been predicted by Jacques Monod and François Jacob and which was thought to be responsible for switching genes off when their role in protein synthesis was not required by a cell. The task of uncovering the identity of this genetic suppressor was extremely difficult and Gilbert's eventual success established his reputation as an accomplished researcher. His next achievement was greater still – and it brought him a Nobel Prize.

Like Sanger, Gilbert was drawn into the search for a way to sequence DNA, the next milestone in suborning the power of the human gene and tying it to the yoke of medical science. But whereas Sanger rejected the principles of biological breakdown which led him to the conquest of the structure of insulin, Gilbert, working with the biologist Allan Maxam, stayed loyal to it. Using a special type of enzyme called a restriction enzyme (which we will discuss in greater detail), he sliced his DNA strands into sequences that were around 100 bases in length. Then he separated them into four portions, each of which was subjected to one of four chemical reactions: one cut the DNA at adenine (A) bases, another at C, another at G, and another at T. Crucially, however, Gilbert allowed this process to work only incompletely, paralleling the approach of Sanger who restricted his nucleotide build-up by using chain terminators. Gilbert was then left, in each portion, with a range of DNA sequences all ending in A in one case, and C, G and T in the others. Once again, by using gel electrophoresis, it was possible to separate these pieces according to size and read off results quite simply.

Today, Sanger's technique is the more widely used of the two. This should not diminish Gilbert's achievements in making sequencing a commonplace technique, however. In any case, we shall soon come across Gilbert again, though first we must look at a couple of other breakthroughs that were required before scientists could turn DNA analysis into a common, practical procedure. In particular, researchers needed to find ways to amplify DNA so that enough could be made available for sequencing. This problem was solved by Stanley Cohen and Herbert Boyer.

In 1969, Boyer, at the University of California, San Francisco, had begun work on restriction enzymes, a special category of proteins used by bacteria to defend themselves against viral attack. A restriction enzyme cleaves an invading virus's DNA at a specific locality, slicing through it with razor-like precision. For instance, it might typically cut at a specific genetic sequence such as CCGTA. Every time the enzyme comes across this phrase it will sever it at a specific point; between the G and the T, say. In this way, it cuts cleanly through the DNA strand and its complementary partner. Over the years, restriction enzymes have become powerful weapons in the molecular biologist's arsenal, but in 1969 researchers were only beginning to appreciate their potential.

The restriction enzymes that interested Boyer come from the *E. coli* bacterium and they have a special property. When these biochemical scissors cut through a double strand of DNA, they do not do so cleanly, leaving blunt ends. Instead, they produce step-like ends. In a rare flash of helpful simplicity, scientists decided to label these 'sticky ends' – and for a good reason. The overlapping flaps created this way have exposed bases on them, and these, of course, attract other complementary bases, because that is how DNA replicates itself. These sticky ends are therefore like tiny, highly specific fingers of Velcro which tend to rejoin other similar ends given half a chance. They are a sort of biological cutter and welder rolled into one.

And the potential of these biochemical fasteners was immediately appreciated by Stanley Cohen when he heard Boyer describe them in a lecture in Hawaii in November 1972. At the time, Cohen was working on plasmids, microscopic circlets of DNA that float outside a bacterium's chromosome (which is just one large DNA molecule), and reproduce independently of it. (Plasmids were discovered by Joshua Lederberg as part of his work on bacterial genetics.) As Cohen listened to Boyer, he realized that any piece of DNA cut with one of Boyer's enzymes could be inserted into a plasmid that had also been cut with the same enzyme – regardless of species. You could

take a gene from a walrus and stick it in a plasmid from a fruit fly, or whatever took your fancy – because Boyer's enzymes produce compatible sticky
ends. Cohen immediately arranged a meeting with Boyer, and in a deli-
catessen near Waikiki beach the pair munched corned beef sandwiches and
agreed on a collaboration that was to have profound consequences not only
for science, but for the worlds of multinational business and medicine as
well.

What Boyer and Cohen went on to do was straightforward, fairly effort-
less – and quite revolutionary. They created a composite plasmid made out
of DNA from two different types of bacteria and inserted this intact into an
E. coli bacterium. The new plasmid happily replicated itself many times in
each bacterial cell, and the bacterium itself divided every twenty minutes. In
this way, countless copies of the new plasmid genome were created in a very
short time. The report of this historic experiment was published in
November 1973 in the *Proceedings of the National Academy of Sciences*. The
day of the scientific cloner had begun.

We should be careful what we mean by the term 'cloning', however. Used
in its original sense, the word implies the creation of an entire organism from
only a scrap of its tissue. After all, each cell in a living being contains all the
DNA that is required to direct the growth and development of the entire
plant or animal. It is a vision that has fascinated many writers and film-mak-
ers and has provided the stimulus for bookshelves of stories in which
countless copies of a person have been 'cloned' from a single one of their
cells. Examples include the dozens of Hitlers who were made in Ira Levin's
book, *The Boys from Brazil*, as well the Leader who was to be recreated from
his nose (his only surviving organ after assassination) in Woody Allen's
science fiction spoof, *Sleeper* – though the most spectacular cinematic example of cloning was provided by Steven Spielberg in his blockbuster *Jurassic
Park*. It is certainly an intriguing idea, though the complexity of the process
by which DNA is switched on and off during cell differentiation means that
it is likely to remain science fiction for some time.

However, in the definition that became fashionable after Boyer and
Cohen, cloning has simply come to mean using living organisms to reproduce endless pieces of DNA. Think of it as a form of biological
photocopying. And an extraordinarily effective type it is as well. Inserted in
a bacterium, a plasmid – engineered to carry an extra foreign gene – can
create a million copies of itself, and that extra rogue gene, in a day.
Combined with the techniques for rapidly sequencing DNA that were then
being pioneered by Sanger and Gilbert, this ability to provide endless strips
of genetic material suddenly allowed molecular biologists to isolate pieces of
DNA, multiply them and read off their sequences. The discrete, inscrutable

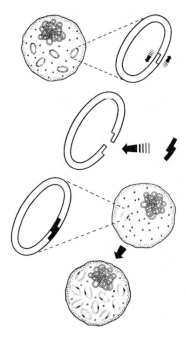

The art of cloning. A plasmid is cut open using a restriction enzyme that leaves over-lapping flaps exposed. A gene with similar flaps is inserted and the plasmid is returned to its bacterial host where it continues to replicate – along with the gene.

gene whose existence had been identified only eighty years previously by Gregor Mendel now lay exposed to the molecular biologists' probing. The black box that controls our heredity had been opened and very soon scientists were rummaging through the contents.

To date, they have made sense of only a few per cent of that material. Yet that small fraction has been enough to alter radically our perception of ourselves, our recent evolution and the way we treat illness. Just how dramatic is that re-evaluation of the human condition can be witnessed in the succeeding chapters of this book. For the moment, however, we can get a taste of these momentous implications by considering the fate of Herbert Boyer. His story, more than any other geneticist's, illustrates starkly how biology has become the dynamic science of the late twentieth century.

Shortly after completing his cloning experiments with Cohen, Boyer was approached by a financier called Robert Swanson, an eager, twenty-eight-year-old venture capitalist (and ex-biochemist) who had the then crazy idea that there were fortunes to be made out of molecular biology. Was it possible, Swanson asked, not just to use gene-splicing technology to create countless identical strips of an entire gene, but to exploit the technique to make that gene's product? In other words, could genetically altered bacteria be persuaded to express proteins foreign to their constitutions, and in particular, could they do so for human ones?

Boyer invited Swanson to his laboratory to discuss the idea, but told him he could only spare twenty minutes. The idea stuck in Boyer's mind, however, and the pair ended up in a local bar for nearly four hours. 'All the academics I called said the commercial application of gene splicing was ten years away. Herb didn't,' said Swanson later on. In other words, Boyer's answer to Swanson's question about the possible manufacture of human proteins using genetically altered micro-organisms was a 'yes', because a plasmid could code for a protein and use the bacterium's internal protein-making machinery to produce it. And in giving this answer, Boyer set in motion forces that were to establish an entire new industry: biotechnology.

Boyer borrowed $500 and joined Swanson in forming a company which would exploit this new science. They called it Genentech, for genetic engineering technology. As Boyer later explained: 'I wasn't happy with Swanson and Boyer, and he (Swanson) was not happy with Boyer and Swanson.' So Genentech it was. Little did the pair know what they had started.

The fledgling company's first target was a modest one, however. It was to create a corporate strategy. In other words, Boyer and Swanson had to decide what human proteins should be top of their pharmaceutical sales catalogue. They chose insulin as their first.

Now this might seem an odd choice. Ever since Frederick Banting and Charles Best, at the University of Toronto, had shown in 1922 that absence of this critical protein was the cause of diabetes, insulin – extracted from cows, pigs and other animals – has been used to prevent glucose from fatally building up in the bloodstream of diabetics. Millions of lives have been saved through the administration of such injections. So why look for a new way to make insulin? The answer is threefold. Firstly, about 10 per cent of diabetics suffer side-reactions to bovine and porcine insulin. Secondly, doctors had discovered that as more and more people adopted Western lifestyles around the globe, more and more were succumbing to diabetes. Some analysts reckoned that insulin could become a scarce commodity as a result. Genentech thought it could fill the void. But most importantly, the company wanted to make a protein with a recognisable name. From the very start, it sought credibility – and insulin should provide that, Boyer and Swanson reckoned.

They decided on a test run, however, and began their first experiments, not on the insulin gene, but on one that codes for a protein called somatostatin. And to do that, Boyer and Swanson called in the aid of two researchers, Arthur Riggs and Keiichi Itakura, the first scientists hired to join the staff of Genentech.

The pair were already experts on somatostatin, a protein that is involved in a cascade of biochemical activity in the body. Its precise function is to inhibit the release of another natural protein, human growth hormone, or HGH. If people fail to make enough somatostatin, the body proceeds to manufacture unlimited amounts of HGH. The result is a rare condition known as gigantism, whose victims grow uncontrollably. In those days, somatostatin could, fortunately, be extracted – though at great expense, and in infinitesimal amounts – from the brains of thousands of freshly slaughtered sheep. Genentech reckoned it could do better and decided to cut its gene-splicing teeth on somatostatin, and for good reason. The protein is tiny, only fourteen amino acids long, and its gene is therefore diminutive as well, small enough to be fairly easily put together from its basic A, C, T, G ingredients. (It is a general rule that large proteins are coded for by large genes, and small proteins by small genes.)

On 15 August 1977, Genentech scientists succeeded in inserting the somatostatin gene inside the genomes of living bacteria which were then persuaded, under the tutelage of this DNA interloper, to begin manufacturing somatostatin. It was a historic moment, for this was the first time that a human protein had ever been generated by a living creature outside the body of a man or woman. But for Genentech it was just the beginning. Insulin was still its prime goal, and nothing had changed the determination of the company's scientists to be the first to clone its gene, and pave the way for the manufacture of the protein in commercial quantities.

It was not a one-horse race, however. Researchers from other departments of the University of California, San Francisco, were also closing in on the insulin gene. In addition, a group from Harvard, led by the ubiquitous and ever-ambitious Walter Gilbert, was homing in on the same genetic target. It proved to be a contest of epic proportions, tense and closely fought, as Stephen Hall records in *Invisible Frontiers*, his splendid, racy account of *The Race to Synthesise a Human Gene*, as the book is subtitled. In particular, the name of Gilbert constantly occupied the minds of the Genentech scientists, his fearsome intellectual reputation pressing the researchers – who now included Dennis Kleid and David Goeddel – to desperate, almost constant work shifts in their drive to clone human insulin.

It was a fraught existence; the young researchers worked in an old San Francisco warehouse, under polythene sheeting for a roof, and with no technical support staff, feverishly battling towards their goal. Then on 9 June 1978 Gilbert revealed he had cloned and expressed insulin in bacteria. First news of the experiment caused utter dismay among the San Francisco scientists. But gloom quickly turned to relief when it was discovered that Gilbert and his team had only cloned rat insulin. The battle for human insulin was still on.

(The prodigious Gilbert did not confine his efforts purely to the insulin race, even at this juncture, however. In 1977 scientists had discovered that not every piece of DNA within a gene actually codes for a protein. There are intervening sequences which do not code, and are spliced or cut out during RNA transcription. These extra sequences have the effect of making a gene much longer than it needs to be to code for a protein, and to this day these stretches of seemingly purposeless DNA pose something of a mystery to molecular biologists. Gilbert gave them their names – exons for the pieces that do code for a protein, and introns for the excised regions. He even went on to speculate elegantly in *Nature* [8 February1978] about the evolutionary origins of these strange intra-genic segments.)

In the end, Genentech got there first. In the early hours of 24 August 1978, its scientists mixed critically balanced amounts of the protein that makes up the A chain of insulin with that of the B chain (see Chapter Three). The two proteins had been made separately by *E. coli* bacteria that had been 'spliced' with the chains' two respective genes, each of these genes having been painstakingly put together from their basic individual A, C, G and T base ingredients. Using radioactive tracers, the researchers showed, to their considerable delight, that small amounts of human insulin had fused together from the two basic components. The hour of the genetic engineer was at hand. Two weeks later, the results of this historic experiment were announced along with news that Genentech had signed a deal with Eli Lilly,

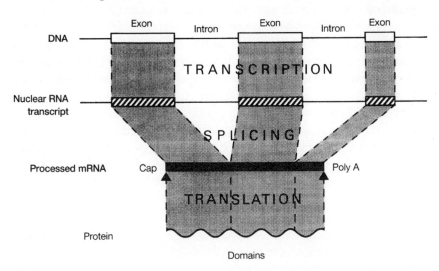

During protein synthesis, all non-coding regions – introns – are cut from messenger RNA. Only the coding sections – exons – remain, and are responsible for the manufacture of amino acid chains.

the pharmaceutical giant which would be responsible for the mass manufacture of human insulin.

The disclosure was made only three years after Swanson had first posed his historic question to Boyer, an astonishingly brief interval and certainly much briefer than had been predicted by other molecular biologists. The inception of research was followed with unprecedented swiftness by the actual manufacture of the first genetically engineered human protein, showing how ripe was the technology and how prescient Boyer had been in his confidence.

Soon a multitude of other new biotechnological companies, mostly based in the United States, began to mushroom; commercial outfits whose expertise lay entirely with the gene-splicing talents of their scientists. Thanks to their molecular biological aptitude, it became possible to insert increasing numbers of different human genes into bacteria and other simple organisms. Dozens of medically valuable substances have since been created by doctoring micro-organisms this way, creating genetic cuckoos that have been nurtured as 'the real thing' and which, in the end, have yielded supplies of many important pharmaceuticals.

One of the first entrepreneurial outfits to take advantage of this technology was called Biogen, and a key founder was – you've guessed it – Walter Gilbert. Unfazed by losing out in the battle for insulin, Gilbert, and Biogen, earmarked other pharmaceutical products that could be made by genesplicing, proteins such as the various forms of interferon which were thought to have possible anti-cancer and anti-viral powers.

Other companies that were set up at this time include Genex, Centocor, Amgen, Chiron and many others, while the product goals included human growth hormone to counteract dwarfism, blood clotting factors to treat haemophilia, tissue plasminogen activator (TPA) to help angina and stroke victims, and others.

But perhaps the most exciting genetically engineered health product that was produced in the 70s and 80s was erythropoietin, or EPO. Erythropoietin is a growth factor, one of a class of chemical messengers that stimulates organs and tissue to carry out various crucial tasks in the body. EPO is made in the kidneys and it stimulates bone marrow to make red blood cells. Without it, we would die. And that is why most patients with kidney failure usually have to have regular blood transfusions – to make up for the red cells lost because of a lack of EPO stimulation. These saved lives, of course, though patients were usually left tired and weak after a few weeks, until they were given a new transfusion. Then Amgen marketed EPO, made in bacteria. Patients' red blood cells rose sharply as their bone marrow, stimulated by EPO injections, increased manufacture, producing a dramatic improvement in their condition. 'Suddenly people got enough energy to hold down jobs and mow their own lawns,' says one of the pioneers of the application of EPO, Dr Christopher Winearls of the Churchill Hospital, Oxford. 'It was spectacular. It utterly changed their lives.'

By 1980, the world of finance was not only taking notice of this fledgling industry, it was positively begging to buy a slice of its biological action. It got its chance on 14 October when Genentech offered 1,100,000 of its shares for sale at $35 each. What followed left Wall Street stunned. Within twenty minutes of trade opening, Genentech stock had jumped to $89 a share. 'Veteran traders had never seen such commotion over an embryonic company, which only has 140 employees, has sold no product to the public and showed a profit for just one year, at a rate of 2 cents per share,' reported *Time* magazine.

So impressed was *Time* with these technological pyrotechnics that it put Boyer on its cover for the issue of 9 March 1981 – a very rare distinction for a scientist. With typical clipped prose, *Time* gave the following description of the first Lord of the Genome: 'In his faded jeans and open leather vest, a can of Budweiser in his hand, he looks just like a leftover from the 1960s. Back then, in fact, he marched regularly in the streets of Berkeley, California, taking part in civil rights and antiwar demonstrations. Despite his casual look, Herbert Wayne Boyer is a millionaire many times over, at least on paper. More important, he is in the forefront of a new breed of scientist-entrepreneurs who are leading gene splicing out of the university laboratory and into the hurly-burly of industry and commerce.'

Certainly, the rewards for Boyer were colossal. With the leap in price of Genentech stock, his 925,000 shares catapulted his wealth, on paper at least, to more than $80 million. Not a bad return for an investment of only $500. And although the company's shares slid back to around $70 a share, he had become, by any calculation, a very rich man. Nor was he the only person to make a fortune. Many other Genentech personnel, such as Kleid and Goeddel, who had been given thousands of shares when they joined the payroll, were also left with unexpected but very welcome wealth on their hands. Never had science been so financially well rewarded.

However, it soon became apparent that the gene-splicing business was not a licence to print money. Investors found that a surprising number of scientific hurdles had to be overcome before they could begin to get returns for their money. For a start there was the delicate process of inserting a gene exactly where you wanted it. In their original work, the Genentech scientists had placed the A and B insulin genes near a large bacterial gene which coded for a heavy protein called beta-galactosidase. Yields of the two human proteins were low, probably because they were aligned to such a large bacterial protein. As the huge protein emerged from its ribosome 'factory', it appeared to fall off, dragging its insulin chains, only half completed. It took many experiments to find a better site, near a small gene known as trp E, before production of insulin was improved. And manoeuvres of this kind have proved to be necessary for many other gene-splicing ventures. Certainly, tricking bacteria to manufacture high yields of a protein should not be thought of as a straightforward matter. Indeed in some cases this process may just not work at all. A protein may not fold up properly in a particular bacterium, for example. In such cases, molecular biologists try to gene-splice animal or human tissue cells instead. And these procedures can be effective, though the process is more cumbersome and expensive.

And then there was the question of scaling up. In contrast to the brash new techniques of the gene splicer, this was a process that depended on some very old expertise, procedures that were ultimately derived from industries such as brewing and bread-making. (The word biotechnology is sometimes employed to describe the use of any biological process in the manufacture of a product, be it bread or insulin. In this book, however, the term refers only to its narrow, post-Genentech meaning, which was defined by the US Office of Technology Assessment in 1991 as being 'the industrial use of DNA, cell fusion and bioprocessing techniques'.) Using these established methods, molecular biologists expected to grow their altered organisms in vats, so they could later cream off rich brews, not of alcohol or bread, but of human proteins. Production is a delicate business, however, and the investment required for this modern harvesting can be colossal. In April 1980, Eli Lilly spent $80 million in building pilot plants for human insulin manufacture. Then three-phased trials, part of the normal process of drug validation, had to begin – the first to test human insulin's safety, the second its efficacy, and the third to compare its effectiveness against other existing treatments. As a result, the world had to wait until 29 October 1982 – six and a half years after Boyer and Swanson's historic meeting – before permission was given for Humulin, the marketing title for human insulin, to be sold by pharmacists.

Human insulin made by gene-splicing has since carved a reasonable share of the world's pharmaceutical market, though it could hardly be called an overpowering commercial success. Although generally linked with fewer allergic reactions among diabetics, its manufacture has proved to be no cheaper than the old methods of purifying pancreatic material from abattoir carcasses. As a result, animal insulin is still commonly used. On the other hand, as a trail-blazer for genetically engineered medicine, human insulin had considerable impact.

Nevertheless the time taken for a product to reach the market – delays caused by purification, efficacy trials and drug safety legislation – did take the edge off gene-splicing's glowing financial halo. By the time Gilbert's Biogen went public in March 1983, investors' craving for biotechnology stock had already cooled. Offered at $23 a share, the flotation slumped. Biogen continued to lose heavily for several years until Gilbert, who had given up his Harvard professorship to run the company full-time, was forced to resign in December 1984. He returned to Harvard.

Since then the commercial reputation of biotechnology has gone through several cycles in which it has peaked and troughed as regularly as a test tube revolving in a centrifuge. It reached highs in 1983 and 1986, and lows in 1984 and 1988, before beginning a steady, impressive rise from 1989 right through to 1991. And although fluctuations in fortunes will doubtless continue, the long-term future for the gene-splicers looks rosy. 'Biotechnology is likely to be the principal scientific driving force for the discovery of new drugs and therapeutic chemical entities as the pharmaceutical industry enters the 21st century,' stated a US Office of Technology Assessment report in 1991.

However, the biotechnology business has been evolving. New companies spring up while older ones are swallowed by major pharmaceutical firms. In 1986, for example, Hybritech was bought by Eli Lilly for $500 million and Genetic Systems was acquired by Bristol Myers for nearly $300 million. And other takeovers have become quite common. However, the biotechnology community was still quite unprepared for the news that emerged on 2 February 1990. Genentech, the enfant terrible of gene-splicing, was to be taken over by Hoffmann–La Roche, who paid a staggering $2.1 billion (then worth about £1.3 billion) for 60 per cent of the biotechnology company's equity.

Despite a vigorous research policy, Genentech had run into financial difficulties over the development of Activase (the brand name given to tissue plasminogen activator (TPA) which helps heart attack victims), which was intended to become its principal product. Delays in gaining approval from the US Federal Drugs Administration, as well as the emergence of low-cost competing products and the raising of scientific questions about Activase's efficacy, damaged Genentech's commercial standing and spurred efforts to find a partner. Eventually, its directors picked Hoffmann–La Roche.

By this time, Genentech was employing a staff of 1,850, and nearly every member held a stock option. Thanks to the Roche takeover, which gave the company the right to buy all redeemable Genentech stock for between $38 to $60 a share, these holdings were transformed into cash windfalls of, on average, $60,000. For his part, Herb Boyer collected an estimated $36 million, a fairly satisfactory compensation for losing one's company.

Another merger that raised eyebrows, despite the increasing frequency of takeovers, was that of Cetus, another San Francisco-based biotechnology 'start-up' from the heady days of the mid-1970s. Cetus fell by the wayside, being swallowed up by another biotechnology firm, Chiron, in 1991, after pinning its faith on a compound called interleukin 2, which was given the brand name Proleukin. The company hoped Proleukin, a treatment for kidney cancer, would be its major moneyspinner. Unfortunately, the US Federal Drugs Administration refused to sanction its use, even though Proleukin had been approved for use in Europe. Cetus was left without a major revenue spinner, its share price plummeted, and it was eventually taken over by Chiron.

What is strange about this story is that Cetus had been a patent-holding developer of perhaps the most exciting piece of molecular biological technology to have been developed in the 1980s – polymerase chain reaction, or PCR. Invented by Kary Mullis, and developed by a team of Cetus scientists led by Henry Ehrlich, PCR can be used to make millions of copies of a single piece of DNA in a test tube without using bacteria, a power that has 'revolutionized molecular biology', to quote the normally restrained journal *Nature*. Certainly, its impact has been extraordinary, transforming the detection of genetic disorders, cancers and infectious diseases as well as paternity and forensic analysis. And it is not hard to see how it has brought about this change. Take the following description of PCR, written by Mullis, which appeared in a special 1991 edition of *Scientific American* published to commemorate 'the century's greatest scientific breakthroughs'.

'Beginning with a single molecule of the genetic material DNA, the PCR can generate 100 billion similar molecules in an afternoon. The reaction is easy to execute: it requires no more than a test tube, a few simple reagents and a source of heat. The DNA sample that one wishes to copy can be pure, or it can be a minute part of an extremely complex mixture of biological materials. The DNA may come from a hospital tissue specimen, from a single human hair, from a drop of dried blood at the scene of a crime, from the tissues of a mummified brain or from a 40,000-year-old woolly mammoth frozen in a glacier.' Short of a magic wand, a molecular biologist could scarcely ask for anything more powerful. It is scarcely surprising that PCR earned the eccentric Kary Mullis a share of the 1993 Nobel Prize for chemistry.

And like so many great scientific ideas, the concept of polymerase chain reaction arose by accident. In 1983, Mullis was working for Cetus, spending most of his time 'puttering around with oligonucleotides,' as he puts it. An oligonucleotide – which is also known as a primer – is a short piece of DNA which, when radioactively labelled, can be used to determine if a sample of genetic material contains a specific sequence of As, Cs, Gs and Ts. Mullis's principal efforts were then directed at improving the ease of use of these oligonucleotides in the sequencing of DNA.

'One Friday evening late in the spring, I was driving to Mendocino County with a chemist friend,' recalls Mullis. 'She was asleep. I liked night driving; every weekend I went north to my cabin and on the way sat still for three hours in the car, my hands occupied, my mind free. On that night I was thinking about my proposed DNA-sequencing experiment.'

Mullis had a notion that he might be able to bracket a selected piece of DNA with two different oligonucleotides, one locked on to one strand, the second to the other. 'By directing one oligonucleotide to each strand of the sample DNA, I could get complementary sequencing information about both strands,' he thought. In fact, he was on the edge of discovering the polymerase chain reaction.

Suddenly, inspiration gripped him. By carefully controlling the way the oligonucleotide primers are attached to the DNA, and by continuously halting the growth of the complementary strands as they spread along the original DNA, and by breaking them apart again, he could make endless copies of the sought-after sequence. 'The idea of repeating a procedure over and over again might have seemed unacceptably dreary,' says Mullis. 'I had been spending a lot of time writing computer programs, however, and had become familiar with reiterative loops – procedures in which a mathematical operation is repeatedly applied to the products of earlier reiterations. That experience had taught me how powerful reiterative exponential growth processes are. The DNA replication procedure I had imagined would be just such a process.' Once again, the cross-fertilization of ideas from one discipline to another induced an utterly new way of looking at a problem.

'Excited, I started running powers of two in my head: 2, 4, 8, 16, 32 . . . I remembered vaguely that two to the tenth power was about 1,000 and that therefore two to the twentieth power was around a million. I stopped the car at a turnout overlooking Anderson Valley. From the glove compartment I pulled a pencil and paper – I needed to check my calculations. Jennifer, my sleepy passenger, objected groggily to the delay and the light but I exclaimed that I had discovered something fantastic. Unimpressed, she went back to sleep. I confirmed that two to the twentieth power was over a million and drove on.' Not since James Watt walked across Glasgow Green one morning in spring 1765 and realized that a secondary steam condenser would transform steam power, an inspiration that set loose the Industrial Revolution, has a single, momentous idea been so well recorded in time and place.

Back at Cetus, Mullis tried to enthuse his colleagues about his idea. Few took him seriously. 'In the past, people had generally thought my ideas about DNA were off the wall, and sometimes after a few days I had agreed with them. But this time I knew I was on to something.' It was not until Mullis was able to demonstrate physically that his idea worked that Cetus realized they were on to a major moneyspinner, though as we now know the company was still unable to exploit properly PCR's potential. As part of its financial struggles, Cetus sold its PCR division to Hoffmann–La Roche before eventually succumbing to the predations of Chiron.

Today, the fruits of Mullis's mountain drive are used in every field of molecular biology and, as we have said, have utterly revolutionized many of these areas. In later chapters, we shall look at some of the most exciting of these applications, for instance at the way PCR has given unparalleled powers to researchers to isolate and sequence DNA from ancient archaeological samples, and to forensic scientists who can now identify criminals from only a few cells of their blood, semen or hair. But before we do, let us look at how PCR actually works in a little more detail. Then we can appreciate just how it has brought about these revolutions.

To detect a piece of DNA by amplifying it, a scientist first has to know the sequences of two small regions on either side of it. These are oligonucleotides, or primers. Each is typically about twenty bases in length, while the maximum size of DNA in between is usually no more than about one thousand base pairs. Polymerase chain reaction – which is sometimes known as DNA amplification – cannot easily multiply sections any greater than this limit.

Now, let us suppose a scientist is trying to identify whether or not an individual's genome carries a stretch of DNA that makes up part of the gene for sickle cell anaemia. In other words, the researcher wants to know: is this person a sickle-trait carrier? To find out, a sample of his or her DNA is placed in a solution along with the two primers that border a chosen part of the sickle cell gene, that section being one that will uniquely identify the mutated haemoglobin gene. In addition, a plentiful amount of DNA building blocks of A, C, G and T bases, as well as the copying enzyme DNA polymerase that was exploited by Sanger in creating his sequencing technology, are added.

All the scientist then has to do is to heat and cool the solution in a regular cycle – it is as simple as that. The heating causes the DNA strands to separate. Then the solution is cooled, allowing the primers to latch on to their specific targets on each of the two strands, and from these points the bases are added, one at a time, by the polymerase enzyme to create a complementary strand. Then the solution is heated again. The two newly created double strands separate, and the process of DNA assembly is repeated when the solution is cooled. 'These cycles can be as short as one or two minutes;

during each cycle the number of DNA target molecules doubles,' says Mullis. (The crucial point is that the DNA building-block bases will not attach themselves unless a primer is first affixed to a DNA strand. It must be there to initiate the growth of the secondary strand. The bases will therefore grow only on the section that is pinpointed by the primers, and not elsewhere.)

It is then a simple matter of placing the DNA on a gel, before passing a current over it to show if pieces that are the length of the hunted sequence are present. The DNA bands are revealed by marking them with a radioactive probe. Once the DNA has been isolated in this way, there are a variety of simple means to find out if it contains the sickle cell mutation. The point about the whole process of PCR is that it gives so much DNA, of exactly the right sequence, that the search becomes extremely easy.

That, then, is the absurdly rudimentary, but quite brilliant, idea that Mullis had on his moonlight drive. (Ironically Gobind Khorana, one of the researchers who played a key role in the unravelling of the genetic code, had suggested such an idea several years earlier but had not acted upon it.) One or two refinements have since been made, of course, such as the use of DNA polymerase extracted from the bacterium *Thermus aquaticus* which lives in hot springs. 'The polymerase that we originally used was easily destroyed by heat, and so more had to be added during each cycle of the reaction,' says Mullis. 'The DNA polymerase of *Thermus aquaticus*, however, is stable and active at high temperatures, which means that it only needs to be added at the beginning of the reaction. This high-temperature polymerase is now produced conventionally by genetically engineered bacteria.'

This last refinement means that a scientist can set up a polymerase chain reaction, go for lunch, and come back to find that twenty repetitions of the reaction will have created a million copies of the DNA stretch which he or she is investigating. A mouthwash of saliva, which typically contains a few cells from the lining of the cheek, provides enough genetic material for a scientist to determine if an individual is a carrier for a disease such as cystic fibrosis, for example. Equally, a hair left at a murder scene, or the crumbling remains of an ancient skeleton, can now be persuaded to reveal their genetic secrets, thanks to PCR.

Dideoxy sequencing, cloning, polymerase chain reaction – these are the tools of 'the new genetics' that were developed in the 1970s and 1980s, technologies that spawned new industries, turned startled researchers into multimillionaires, and opened up the human genome to minute investigation. Its impact has been astounding, as Robert Weinberg, a leading US cancer researcher, made clear in the very issue of *Scientific American* in which Mullis outlined the discovery of PCR. 'The newly gained ability to describe and manipulate molecules means the biologist is no longer confined to

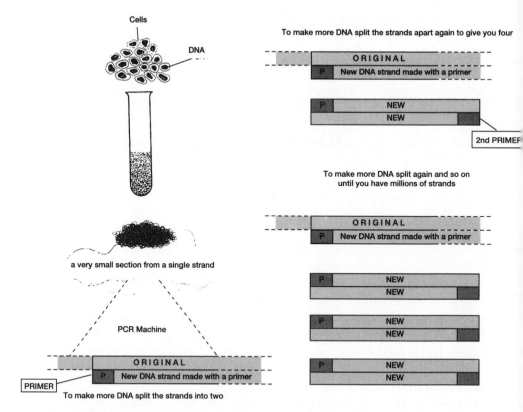

The ultimate gene machine. To carry out a polymerase chain reaction, a small section of DNA is removed from a cell. Then its two strands are separated, and primers attached. Complementary strands attach themselves to the original. Then the two new DNA sections are separated, and the process repeated. Each time this is done, the number of copies of the DNA section is doubled.

studying life as the end product of two billion and more years of evolution,' states Weinberg. 'The new technology has made it possible to change critical elements of the biological blueprint at will and in so doing to create versions of life that were never anticipated by natural evolution. In the long run this may prove to be the most radical change deriving from the power to manipulate biological molecules.'

The rest of this book will show how spectacular have been the consequences – for cancer research, immunology, archaeology and forensics – of this manipulation. But first let us investigate what has been the impact on those families who are afflicted by hereditary diseases, ailments which sometimes seem to strike out of the blue, and in other cases dog dynasties like a curse. Today, these once impenetrable illnesses have been exposed to the relentless probing of the molecular biologist.

5

Cutting the Cord

Woody Guthrie lived a peripatetic existence that has acquired, over the decades, the romantic resonance of a legend. For years he rode freight trains across America, living with the downtrodden of the Depression, and gaining experiences that later inspired some of his finest songs; 'This Land is Your Land', 'This Train is Bound for Glory', and many others. Guthrie became one of America's greatest folk singers and writers, and his influence spanned a generation of musicians from Pete Seeger to Bob Dylan. He was an extraordinary individual – with an extraordinary past.

And the most disturbing part of that history was bequeathed to him through his mother's family. Nora Guthrie gave birth to Woody on 14 July 1912 and raised him at the family home in Oklahoma. His early life was happy but was slowly transformed by his mother's increasingly odd behaviour. She began to neglect her children, leaving them unwashed and ill-clad. Then she became violent, her temper erupting without warning. 'Her face would twitch and her lips would snarl and her teeth would show,' Woody once wrote. 'She would start out in a low grumbling voice and gradually get to talking as loud as her throat could stand it; and her arms would draw up at her sides, then behind her back and swing in all kinds of curves.'

Then came tragedy. Woody's elder sister Clara was killed in a mysterious fire at the family home. Neighbours blamed Nora for the death, though no proof was available. Then in 1927 Woody's father Charley was badly burned in another blaze. This time it was clear that Nora was the culprit. It was the last calamitous act in the Guthries' disintegrating domestic life. Charley was taken to hospital, Nora was institutionalized in a mental home, and Woody and his brothers and sisters were simply abandoned. Woody spent his adolescence and early adulthood begging, playing songs in the street and doing

odd jobs. He only visited his mother once, as Joe Klein recalls in his definitive biography, *Woody Guthrie: A Life* (Knopf, 1980). 'She didn't recognise
him. The doctors said she was suffering from something called Huntington's
chorea, a nervous disease that couldn't be cured and only got worse, and
they told him other things too, but he wasn't listening very carefully.'

Woody went on to make his career, carving out a tempestuous trail of
wrecked relationships in the process. Then in 1951 his marriage to his second wife, Marjorie, took a desperate turn. He became wildly jealous and
violent. Like a ghastly nightmare that cannot be shrugged off, he began to
mimic his mother's deranged behaviour. His movements became unco-ordi-
nated, he suffered dizzy spells and bouts of black melancholy. Woody was
eventually hospitalized and in September 1952 he was diagnosed as having
'psychosis associated with organic changes in the nervous system associated
with Huntington's chorea'.

Over the next fifteen years, Woody's condition declined inexorably. He
displayed the condition's strange dance-like jerky movements of hands and
arms, and progressed through stages of worsening emaciation, exhaustion
and dementia, all the time his limbs flailing uncontrollably. At the end,
Woody weighed less than 100 pounds. 'His skin was waxy and translucent,
his bones sticking through it,' says Klein. 'He was so weak that he didn't even
have the energy to shake very much. His movements were constant, but
almost gentle now.'

On 3 October 1967, Woody Guthrie died, a victim of an ailment that has
been labelled as 'the most demonic of diseases'. It is a just description. Apart
from the terrible effects on those who inherit Huntington's deadly gene,
there was, before molecular genetics stepped into the fray, no way to tell if a
person would succumb until late in middle age. As a result, many victims
have children before they have been properly diagnosed. A healthy husband
or wife is then left to watch as spouse and children sicken and die.

In the past, the tragedy was compounded because the disease's symptoms
were often mistaken for drunkenness or madness. Sometimes a person died
because their physical control had been destroyed; a proper diagnosis could
not then be made. In such cases, sons and daughters were left unaware of
their lethal heritage and the disease was unwittingly passed on through generations. In Woody Guthrie's case, for example, his maternal grandfather was
killed in a mysterious horse-riding accident. His children therefore had no
portent of what lay ahead. And Woody's only warning about the disease was
given when he made his single visit to his institutionalized mother. An un-
educated lad, shocked at her condition, he simply did not take in the
doctors' words. Woody Guthrie fathered six recorded offspring, of whom at
least two – Gwen and Sue, daughters by his first marriage – had begun to

display the symptoms of the disease not long after Woody died.

Huntington's chorea is caused by a single mutant gene which affects about 5 in every 100,000 births, creating an unforgiving progression that stretches from the distant past to the future, linking victims in a seemingly unbreakable web. But with the death of Woody Guthrie, this despairing picture was to change. Thanks to work initiated by Marjorie, his widow, it is now possible to cut through the cord that has, up until now, tied families to the disease.

And as you might expect, the techniques that have brought about this liberation are the work of molecular biologists, in breakthroughs based on developments that we have discussed in the first four chapters. It is now time to look at the direct consequences of those advances and examine how they changed the lives of men and women caught in the once hopeless snare of inherited disease. Today, a host of different ailments – some trivial, most life-threatening, but each caused by the effect of a single mutated gene – have succumbed to advances in molecular biology, progress that has been made in one glittering decade of genetic achievement. We shall now look at some of the most important of these, starting with the 'demonic' illness that killed Woody Guthrie.

Huntington's chorea is probably a very ancient disease. It was only identified relatively recently, however, being described in 1872 in the only scientific paper published by George Huntington, the physician son, and grandson, of physicians. One of Huntington's earliest memories was of riding with his father on his rounds in Long Island when they came upon 'two women, both tall, thin, almost cadaverous, both bowing, twisting, grimacing'. The memory haunted him, and after graduating from medical school Huntington decided to write a paper on his observations, using notes made by his father and grandfather as well. He called the disease a chorea because of its dance-like movements (chorea comes from the same root as choreography).

Huntington had access to knowledge about several generations of shakers (as he termed victims), and so he was able to make the following reasonably accurate description of the disease's behaviour: 'When either or both parents have shown manifestations of the disease . . . one or more of the offspring almost invariably suffer from [it], if they live to adult age. But if by any chance these children go through life without it, the thread is broken and the grandchildren and the great-grandchildren of the original shakers may rest assured that they are free from the disease.'

The disease, in short, is caused by a dominant gene, one that behaves in the same way as those involved with traits such as the ability to taste that strange chemical, phenylthiocarbomide (PTC), which we encountered in

Chapter One. Other examples of dominant genes are those which dispose humans to have brown eyes and pea plants to have round seeds. Not surprisingly then, Huntington's chorea (or Huntington's disease as it sometimes known today) is called a dominant, inherited illness. It is caused by a single faulty gene which dominates its normal counterpart. There are usually no symptomless carriers, and as an affected parent can give only one of two genes to his or her offspring, one defective and one normal, this means each of their children will have a 50-50 chance of being given the gene.

Using Huntington's explanation of how the disease passes through generations, it was then possible to outline its grim course over the centuries. Researchers found that several hundred chorea victims from New England could be traced back to four families who arrived in the area in the seventeenth century. One of these families was then shown to have come from pilgrims from the English village of Bures, in Suffolk, in 1630. Intriguingly, several women descended from this family were burned at the stake during the Salem witch trials in 1693, implying that their symptoms – violent tempers and flailing limbs – may have been interpreted as those of demonic possession. Certainly, the frightening aspect of Huntington's victims suggests that in the past societies may have treated these poor people very harshly.

However, the real trouble was that even though Huntington had uncovered the disease's basic pattern of inheritance, scientists were left no further forward in understanding what caused the illness. No clue was ever found to explain how the defective gene causes the disease, nor was there any idea what its normal counterpart does. It was not possible to exploit the approach of standard genetics which has been used with other inherited diseases such as sickle cell anaemia. In these cases, a protein defect – such as an abnormal oxygen-carrying haemoglobin – has been identified, and linked to a genetic lesion that has revealed where the gene for the normal should be. For Huntington's chorea, as with many other inherited diseases, that approach has been impossible because of the byzantine nature of the ailment's symptoms. The chemical nature of the disease remained a mystery, and so did its gene. All that seemed clear was that the mutated gene must make a protein that is subtly altered from its normal counterpart, turning it into poison of some sort. Perhaps this toxin slowly accumulates in the nervous system to produce the terrifying battery of symptoms: derangement, depression, violence, and those awful relentless jerking limbs that, in the end, leave victims in states of terminal exhaustion.

After Woody's death, Marjorie decided that she had to try to do something about the disease. So she set up a Committee to Combat Huntington's Chorea. At the time, the late 1960s, its prospects looked forlorn. There were still no clues about the disease's biochemical roots, nor was there any sign that treatments were about to be developed. Yet Marjorie's committee was eventually to produce quite monumental results. Her first real break came when she encountered a Californian psychiatrist called Milton Wexler, and one of his two daughters, Nancy. Milton's wife, Leonore, had been diagnosed as having Huntington's chorea, which meant that Nancy, and her sister Alice, each had a 50-50 chance of succumbing later in life. It was a chilling prospect, which had wrecked the lives of many young adults. The Wexlers decided to fight, however, and set up a California chapter of Marjorie's committee. Milton, Marjorie and Nancy made a forceful trio. They lobbied Congress, staged fund-raising rallies, and held workshops to discuss the plight of victims and to mull over proposed remedies.

By 1976, they had managed to persuade Congress to pass a bill creating a federal Huntington's Disease Commission. Marjorie became its chairman, Milton its vice-chairman for research, and Nancy its executive director. The first act of the commission was to hold public meetings to establish the depth of the problem, and to give victims and their families a chance to air their grievances. What the commission heard was shocking, as Nancy recalls. 'We heard of people who had spent their life savings trying to get proper diagnosis of the disease; of a seventy-six-year-old woman, with no social security, who had to look after middle-aged sons who were so badly affected they had to wear nappies all the time; of people visiting relatives in psychiatric hospitals, hearing them screaming and seeing them tied up; of one woman who spent $26,000 on medical bills for thirty-one different doctors before anyone recognised her condition; of families that had been decimated by Huntington's; and of men and women who lost jobs because they were thought be drunk. It certainly put Huntington's in context.'

But to balance these grim findings, the commission also made one extremely important discovery. It uncovered the world's largest collection of Huntington's victims, in a squalid community on the shores of Lake Maracaibo in Venezuela. There, scores of sufferers were found living in interbred families in primitive tin shacks and lakeside huts on stilts – all descendants of Maria Concepcion Sota, who brought the disease to the area in the 1860s. As Nancy put it: 'From one woman, a huge pyramid of suffering has been stretching out over the decades.'

The Maracaibo discovery was a genetic treasure trove. For one thing, scientists who studied these people would know they were almost certainly studying the effect of a single gene. Elsewhere, they could not be sure whether mutations on two or more different genes might be causing the disease – as we now know is the case with Alzheimer's disease, a degenerative illness of the central nervous system which we shall discuss in Chapter Eight. No such problem was likely to be encountered in the tight-knit community round Maracaibo. In addition, the large numbers of related victims would provide a great deal of useful information that would make researchers' work easy. It was even possible that they might find a homozygote Huntington's victim – in other words, a person with two mutated genes. (A person with a single affected gene, and a normal counterpart, is called a heterozygote.) Such individuals might display distinctive symptoms that could aid the scientific search for the Huntington's gene.

In 1981, Nancy led an expedition to Maracaibo and began the painstaking task of collecting blood and tissue samples and unravelling the convoluted relations of the local people. First the researchers prepared the family trees of more than 3,000 individuals. Then the team – neurologists, geneticists, a nurse, a photographer and an anthropologist – started the business of diagnosis, an easy task in the disease's late stages, but much trickier in its early phase. Using subtle motor and intellectual impairment tests, the scientists were able to pinpoint many victims, however.

Then they turned to the problem of collecting blood samples. These had to be fresh and sent to America within forty-eight hours of collection, so Nancy organized a draw day. But the Maracaibo people were suspicious; many thought they would lose precious body fluids and strength if they gave blood. So Nancy organized a draw-day party. 'We gave them Coke and cakes. It was weird. Many had obvious signs of chorea – they were writhing and twisting their limbs. In the end, we had a room of undulating people. Just the same, many of them were still very cautious. They thought we were looking at them as freaks, that they were the only ones in the world to be so cursed. Then I told them that my mother had died of "el mal", as they called the disease. They were stunned. They couldn't believe that America – a country that had put a man on the moon – could be afflicted with a disease like that. Tissue collecting became a lot easier after that.'

Nancy's team eventually assembled 570 samples from sufferers and their relatives. Together with records of their family trees, these specimens were sent to James Gusella, a young researcher at the Massachusetts General Hospital. And that is when the fancy molecular genetics began. Gusella was one of the first molecular biologists to use restriction enzymes for tracking genes through families. This idea had first been put forward by Walter Bodmer and Ellen Solomon in early 1979 in the *Lancet* and had been followed up and amplified by Massachusetts Institute for Technology researcher David Botstein and his colleagues.

The idea is based on the observation that genes that lie close to each other on the same chromosome are often inherited together, a tendency that has proved helpful in the past when trying to pinpoint the sources of some diseases. For example, haemophilia and one form of colour blindness can be inherited together, as we discussed in Chapter Two, because the genes lie near each other on the X chromosome. The gene for haemophilia therefore acts as a flag for the colour blindness gene, and vice versa. In other words, diseases are sometimes co-inherited in some families. The symptoms of one, therefore, act as markers for the other ailment. But until the development of modern molecular biological techniques, it was very difficult to make much use of genetic markers because there were so few of them. Restriction enzymes changed that.

Scientists realized that by choosing exactly the right restriction enzyme, they could create a set of DNA fragments each with recognized distinctive differences that could be used to differentiate individuals. They could use these differences to track genes – such as those for single inherited illnesses – through families. The consequences would be profound. Such a genetic banner would tell if a person carried an affected gene because they shared a marker with an affected parent. It would also tell scientists exactly where on a chromosome the disease must lie, because in knowing where the marker came from, they would know where the gene lay.

Now this is a highly simplified explanation, so let us look at gene tracking in a little more detail. But before we do, we should consider the issue of variation in human DNA. If we compare any two chromosomes in regions that lie outside genes, in the non-coding sections, there will probably be a one-nucleotide difference for about every 250 nucleotides of chromosome. That is a measure of how much our genomes vary in make-up. And these variations can be used to flag the presence of a defective gene.

To do this, the molecular biologist adds a restriction enzyme to a person's DNA, which is shredded into pieces as a result. But a cutting site on one chromosome may be absent, or an extra one may be present, when compared with another chromosome, because of the genetic variation that we have just discussed. As a result, the restriction enzyme will break apart some of the DNA of the two chromosomes into different-sized fragments. These distinct pieces are known as restriction fragment length polymorphisms, or rflp, or more colloquially 'riflips'. (The word polymorphism means, roughly, 'different shapes'.)

And what a scientist seeks are polymorphisms that lie near a defective gene so he or she can use them to track that mutation through a family. What is desired is a distinctive strip of DNA that appears only in the genomes of victims, and a different-sized fragment that is associated only with family members free of a given inherited disease.

To find these telltale genetic signals, the researcher puts the DNA fragments created by the restriction enzyme through an electrophoresis gel. The different pieces are then graded for length, as we saw in the previous chapter.

The next step is to find out if one of these segments is associated with a particular gene. To do this, a Southern blot is made of the pieces. This rather odd-sounding technique is named after its inventor Ed Southern of Edinburgh University, and it involves transferring the DNA portions to a special paper to which it is possible to add a radioactive probe. The probe will stick to one section of the DNA, the segment that corresponds with its complementary strand. The probe can then be made to reveal its presence on an X-ray film.

Now what the molecular biologist seeks from these procedures is a restriction enzyme that cuts a person's DNA in a way that creates a marker for a disease, and a probe that will pinpoint its position. The first technique creates polymorphisms, genetic segments of different sizes that are graded for size by the electrophoresis. In the second process, the probe reveals the marker's presence, hopefully in a way that varies directly with the progression of a disease gene as it passes through a family.

And that is what Jim Gusella did with Huntington's chorea. He split up the DNA from victims and relatives using a restriction enzyme, and then he added his probes. He expected it would take hundreds of probes, each from a different part of the genome, to highlight a 'riflip' variation that was co-inherited with the disease. He got as far as his twelfth probe, codenamed G8. It seemed to link with Huntington's chorea victims from Maracaibo families. So Gusella tried again, using a bigger sample of people this time. This time his probe unequivocally associated with the disease's progress through the families. Gusella had found a genetic marker for Huntington's chorea almost before he had had time to get down to serious molecular biology. It was an extraordinary piece of good fortune and as a result Gusella ended up with the nickname Lucky Jim; not that he cared.

What Gusella had found was a piece of DNA that came in four varieties, A, B, C and D, of which version C was inherited 19 times out 20 along with symptoms of Huntington's chorea among the Venezuelan families. The fact that the marker was nearly always co-inherited and was not usually separated from the disease gene during chromosome reshuffling at meiosis suggested that they nestled close together. And the G8 gave another piece of information. The marker and, therefore, the gene that causes Huntington's chorea lay near the end of the short arm of chromosome 4. Nancy was overjoyed when she heard the news, cavorting round her office, yelling, 'We've found the gene, we've found the gene.' Gusella, Wexler and the rest of the team published their report in *Nature* in November 1983.

The consequences of the discovery were profound. Firstly it gave scientists a handle for pinpointing the disease gene – its exact location, structure, and finally the mutated protein that it produces, and the normal version that it should make. With these uncovered, it would be possible to begin designing drugs and other treatments that might alleviate, or even correct, a disease that even in modern times was thought to be a death sentence. That would take time, of course. Indeed it took a depressing amount of time. Despite Gusella's lucky break, scientists were left, ten years later, still seeking closer markers and candidate genes that might be responsible for the disease. As we shall see later in this chapter, other inherited ailments have succumbed much more quickly to the onslaught of the molecular biologist. 'The trouble was that the gene for Huntington's chorea lay at the very tip of a chromosome and DNA behaves very strangely in these regions,' says Dr Hans Lehrach, of the Imperial Cancer Research Fund. 'Genes for other single gene ailments have been found near the centres of chromosomes where DNA behaves in a far more easily understood way.'

But there was an almost immediate implication from Gusella's discovery. Armed with a genetic marker, scientists were able to make early diagnoses of Huntington's chorea. By finding which form of the marker – A, B, C or D – linked with the disease gene in a particular family, it became possible to tell if someone would later succumb to its symptoms. It also became possible to carry out ante-natal screening to determine if an unborn child carried the gene, early enough to carry out a termination if it did. The trouble was that neither procedure can be carried out with absolute precision because of that 1 in 20 chance that the marker will not be inherited with the gene, because of chromosomal reshuffling. Nevertheless, it raised the accuracy of diagnosis from 50 to 95 per cent.

This was clearly an improvement; but it generated serious dilemmas as well. Would insurance companies pressure young men and women from Huntington's families to take the genetic marker test before providing them with cover? Would the spouse of such an individual demand tests before having children? And what would it mean to get a positive diagnosis – in place of merely facing a risk – in young adulthood? These are very important ethical problems which we shall discuss at length in Chapter Thirteen. Suffice to say that a screening test without the corresponding prospect of a cure is not an unalloyed blessing. However, in the long term, the outlook for Huntington's chorea must be viewed as being very hopeful.

In the end, the Huntington's chorea gene was found – in March 1993 – thanks to the joint efforts of teams from Gusella's laboratory, the Massachusetts General Hospital, the Medical School, Cardiff, and the Imperial Cancer Research Fund in London. The disease mutation turned out to be an unstable segment of DNA in a previously unknown gene which codes for a protein also of unknown function. The scientists found that in people without the disease, this segment is made up of between 11 and 34 repeats of the CAG triplet. In Huntington's victims, the segment is made up of between 42 and 100 repetitions.

How this 'stuttering' – which we shall come across in greater detail in Chapter Eight – causes the disease's symptoms, researchers still do not know. However, the crucial breakthrough has now been made. In a very short time indeed a disease that once seemed to be an impenetrable, irreparable biological burden has been forced to disclose its cryptic content. Very soon, the cord that tied Woody Guthrie, and so many others, to their pasts will be severed.

In addition, we should not forget that the problems that beset Huntington's chorea are special ones, particularly when compared with most other commonly occurring, congenital, single-gene ailments. There is the elusive position of its gene; the fact that it is a dominant inherited ailment, which means carriers and victims are one and the same; and the issue of its late onset, associated with confused diagnoses, which has resulted in people unwittingly passing on their condition to children they might otherwise have chosen not to have.

So let us look at an inherited disease to which none of the above applies, but which has been a major source of misery just the same. The effects of Duchenne muscular dystrophy are also calamitous. Passed through symptomless carrier females, it afflicts boys with a deadly molecular cargo that induces a progressive wasting disease of the voluntary muscles.

In its behaviour, but not its symptoms, Duchenne – named after Duchenne of Boulogne who first described it in 1868 – acts like the bleeding disease haemophilia which Queen Victoria gave to her son Leopold, and also passed on, in the form of carrier status, to at least two of her daughters, Beatrice and Alice. And this elusive passage through the generations gives a very precise clue as to the basis of Duchenne, as well as haemophilia. It is caused by a mutation on a X chromosome. How do geneticists know this? Well, they are aware that a child who inherits two X chromosomes will be female, while one who has only one X chromosome plus a Y will be male. This means that for a girl who inherits a failure to make an important protein on an X chromosome, there will be another normal X in her cells to compensate for the troublesome one. It will make the protein and she will display no symptoms. A boy, lacking a second X chromosome, will have no such fail-safe mechanism and will be affected by the symptoms of his single X chromosome's deficiency. Such an explanation precisely accounts for the behaviour of haemophilia which we discussed in Chapter One, and also describes the behaviour of Duchenne, one of the most common and most serious diseases of the X chromosome, a class of illnesses commonly called X-linked disorders.

Duchenne is a remorseless destroyer of muscle strength. Consider this description by Marina Cantouzino of an afflicted brother, four years her junior. 'By the age of ten my brother was unable to climb the stairs and could scarcely pull himself up from out of a chair,' she wrote in the *Guardian*, in June 1984. 'And when his legs began suddenly and frequently to collapse beneath him like those of a young colt, the wheelchair finally entered our lives.' Marina's brother died while still young, a typical fate for victims of the disease. In general, boys with Duchenne are afflicted with such severe muscular wasting in their teens that they suffer relentless contractions of the abdomen and chest. Respiratory failure follows.

Armed with the knowledge that the lethal Duchenne mutation lay on the X chromosome, scientists had a head start in the hunt for the disease's causative gene. But by 1979 they still had no idea where on the chromosome it lay. 'The crucial development was the publication of Bodmer and Solomon's letter in the *Lancet* on genetic mapping,' says Professor Bob Williamson of St Mary's Hospital, London, who, with Dr Kay Davies, made the first breakthrough in our understanding of Duchenne muscular dystrophy. 'Their paper suggested that you could use riflips to map the whole genome. In other words, you could study a gene about which you knew nothing at all. It was a big intellectual leap. Everyone at that time was stuck on the idea that you had to study a protein's chemistry to find its gene. Walter and Ellen showed that you could use DNA sequences and nothing else to define a gene even if you hadn't a clue what protein it makes. All you have to do is use your head and make a few Southern blots. It is wonderfully simple.'

Within a few months, Kay Davies and Bob Williamson handed in their first grant application based on the use of restriction mapping. And they picked Duchenne as their target ailment. They could have selected any one of a cartful of inherited illnesses to test out Bodmer and Solomon's theory, but they chose Duchenne – and for a good reason. They had a crucial advantage in their search, because it was already known on which chromosome the mutated protein resided. 'It was a perfect test bed for the other inherited illnesses for which we had no chromosome clues whatsoever,' adds Professor Williamson.

In October 1979 Ray White and Arlene Wyman, both of the University of Massachusetts medical school, confirmed that restriction-fragment-length polymorphisms could be created and used to track through families. 'That was the final inspiration,' says Kay Davies, now professor and director of research at the Clinical Sciences Centre at Hammersmith Hospital in London. 'White and Wyman showed that restriction mapping was a practical idea.' So Davies and Williamson set about the X chromosomes of

Duchenne families with their riflips and their probes, looking for a variation that was inherited only with the disease. In 1981, a few months after beginning their research, they struck gold. They found a piece of DNA on the Xp21 region on the chromosome's short arm that came in two forms. (Chromosome short and long arms are known as p and q for simplicity, and their characteristic banded regions, which are revealed by special stains, are numbered from the centre to the tip.) One tracked through unaffected individuals in a large Duchenne family, the other linked to victims. They had pinpointed the gene. It was the first successful use of restriction polymorphisms to track down the source of an inherited disease. 'We were absolutely elated,' recalls Davies. The researchers announced their discovery, and a couple of months later came news of Gusella's triumph with his G8 probe for Huntington's. Gene mapping had begun in earnest.

It was enough for Professor Williamson. He had other fish to fry, as we shall see in a few pages, though he still kept a keen interest in Duchenne. Professor Davies, who was soon to move to Oxford, ploughed and within a year she had found a second marker for Duchenne, this one on the other side of the gene from her first marker. The Duchenne gene had been boxed in.

By now many other molecular biologists were also working on Duchenne, and two in particular, Dr Ronal Worton of the Hospital for Sick Children, Toronto, and a group led by Dr Louis Kunkel, of the Children's Hospital, Boston, had launched their own programmes. Slowly and purposefully the trio, Davies, Kunkel and Worton, moved in on the Duchenne gene. In 1986 came the next major development, from a boy who was discovered to have part of his X chromosome missing. This deleted section not only left the poor child afflicted with Duchenne, but also chronic granulomatous disease, a crippling ailment of the immune system, and retinitis pigmentosa, a degeneration of the retina that leads to blindness.

'We were particularly interested because the boy suffered from Duchenne – so we had to assume that the normal gene was on the missing section,' says Dr Kunkel. Investigating this portion led the team directly to an ingenious experiment, and the Duchenne gene. Kunkel and his researchers broke up the DNA of the boy's X chromosomes with a restriction enzyme, and mixed these fragments with DNA from a normal X chromosome. The chromosome pieces that matched each other stuck together. A few did not, and these, Kunkel reasoned, were probably pieces of normal X chromosome with no matching partner. They must come from the slice deleted from the boy's X chromosome. But which piece corresponded to the Duchenne gene? To find out, Kunkel simply removed them and tested to see if any would latch on to the X chromosomes of other Duchenne victims. They all did – except one. The probe was codenamed PERT, and it was, Kunkel realized, one of the pieces deleted from the X chromosome of the boy and which was also missing in other Duchenne children. In short it was part of the Duchenne gene itself.

It was the handle everyone needed. Within a year all three teams were sequencing the Duchenne gene. And what a giant the gene turned out to be. Months later, molecular biologists were still picking out segments of it. 'We were absolutely amazed, and I mean amazed,' recalls Professor Davies. 'We just couldn't believe the way bits of the gene kept popping up all over the X chromosome. It spread itself over a couple of millions of base pairs of DNA.'

Now it is important to remember that not every one of these two million base pairs is part of the gene coding sequence. As we saw in the previous chapter, there are parts of a gene that are sliced out during coding. These bits are snipped out and dropped on the cutting-room floor during the business of gene editing when DNA is turned into messenger RNA. As we saw in the previous chapter, the lost pieces are called introns. The surviving, coding sequences are labelled exons. What surprised Davies, Kunkel and Worton were the oceans of introns that surrounded the few islands of coding exons in the Duchenne gene.

Having found a huge gene, it was expected that it would code for a very large protein, which proved to be the case. The dystrophy protein, dystrophin, as it was named, turns out to be made up of several thousand amino acids, compared to the hundreds that make up most proteins. Researchers have since found that dystrophin plays a critical role in helping muscle fibres realign after they have been stretched out of place. Without dystrophin, muscles lose structure and function and eventually they degenerate. (Sometimes there is a less serious mutation, one that still permits the coding of dystrophin, but in an altered state. Then muscle function is impaired, but not destroyed. Patients are weakened but not fatally crippled. This is Becker's muscular dystrophy, a milder disease first diagnosed in the 1950s by the German neurologist, P.E. Becker.)

In a few years our picture of this grim disease was utterly transformed. But what were the consequences? The first was simple, direct and effective; almost as soon as the Xp21 probe was discovered, screening of unborn children of Duchenne carrier mothers was started. 'Up till then, these women would have to wait till they were seventeen or more weeks pregnant, have an amniocentesis test to determine if they were carrying a boy and if so, elect to terminate,' says Professor Davies. 'It was very distressing for them.' The situation was particularly poignant because a boy has only a 50-50 chance of picking up his mother's affected Duchenne X chromosome. He could just as easily inherit her normal version and so be spared the disease, but without a probe for Duchenne there was no way to tell. The Xp21 marker changed that.

'One of the first cases we had was a woman who was a carrier,' adds Professor Davies. 'She desperately wanted a family, but her first three pregnancies had turned out to be boys. She terminated each one. Then, on her fourth pregnancy, she found she was carrying twins – a boy and a girl. That is when she came to us. We showed that neither child had inherited her affected Duchenne X chromosome. It was a very happy moment for us all.'

Since then, births of Duchenne victims have declined dramatically thanks to the extremely accurate ante-natal tests that are now available. However, Duchenne cases will never disappear, for a very simple reason. About a third of them are due, not to unlucky inheritance, but to spontaneous mutations – which usually occur in the production of sperm cells – that will, of course, never be picked up by screening mothers from families at risk of Duchenne.

As for treatment and cures, these are still being worked on. The trouble is that dystrophin is such a very large molecule, and so will be difficult to move around the bodies of victims in attempts to get it into muscle tissue. Various approaches are being studied, though. One promising prospect concerns the discovery of utrophin, a protein coded for on chromosome 6, which is very similar in structure and function to dystrophin, except that its gene is usually switched off after foetal development. Researchers are now studying ways of switching it back on, with the aim of replacing the missing dystrophin.

It is a dramatic idea, regardless of whether it succeeds or not, for it was proposed less than a decade after the discovery of Xp21. Before then researchers did not even know where on the X chromosome the Duchenne gene lay. Now they are thinking about correcting its mutated effects. We can therefore see that Duchenne has been a singular triumph for molecular genetics. But there was, after all, that crucial head start, provided by standard genetics. Thanks to it, scientists were able narrow down the molecular biologists' options to one single chromosome, the X. So let us look at a third condition, one for which such guidance was not available and which, conveniently, completes our trilogy of types of inherited ailments. We have discussed dominant and X-linked diseases; now we should look at one example of the last but most common category, recessive disorders. It is known as cystic fibrosis, the most frequent inherited ailment to affect Caucasians, a wasting affliction of the metabolism that can strike without warning and with devastating consequences. Consider the following case.

John and Sheila Rose are a professional couple in their thirties from north London. Neither was aware that they carried the gene for cystic fibrosis, until

Sheila gave birth to their first child in 1989. From the start it was obvious that things were badly awry. Little Amanda needed surgery for bowel blockage within two days, and has since lived on a daily regime typical of a cystic fibrosis patient: physiotherapy to prevent liquid build-up in the lungs, antibiotics, and enzyme supplements to aid digestion. 'We were overwhelmed,' says Sheila. 'We had no warning, no notion we were carriers. I was so angry for Amanda at first. Why did it have to be her? Today I am just sad for her.'

Amanda's condition is the result of inheriting two cystic fibrosis genes, one from Sheila and one from John. The disease arrived without warning because neither her mother nor father had ever displayed symptoms, and each was therefore unaware of their genetic status. And that is because they had a normal gene to compensate for the mutated cystic fibrosis counterpart that they carried. In other words, cystic fibrosis is a recessive, inherited illness, caused by a single faulty gene, and Sheila and John are symptomless carriers of that gene.

(It is worth emphasizing at this point the exact difference between a recessive and a dominant disease. In the latter case, a mutant gene is doing something that actively harms the body, as we saw with Huntington's chorea. One gene, on its own, is therefore sufficient to do serious damage, even when its partner gene is normal. That is how the gene 'dominates'. But in the case of a recessive disease, the responsible gene is failing to make a crucial protein. In carriers, one gene makes enough of this protein for normal function, and the other makes none. Such is the plasticity of the human frame that these carriers usually get by on a half dose of protein. Only when a person gets two faulty genes are they in a position in which no functional protein is being made. Then they suffer from symptoms due to the lack of that protein.)

Cystic fibrosis 'behaves' in just the same way as those peas of Mendel, with their recessively inherited wrinkled seeds which we discussed in Chapter Two. In other words, the chance of an affected child (or a wrinkled seed) being produced by two carriers is one in four. The child could have two normal genes, one from each parent; a normal from the father, and a defective from the mother; a defective from the father, and a normal from the mother; or two defectives, one from each parent. Only in the last case will the child be affected by cystic fibrosis, as was the case with Amanda. And when that occurs, a variety of life-threatening symptoms quickly become apparent. Mucous secretions become abnormally sticky, there is progressive lung disease, and pancreatic tissue is destroyed.

In the West, one in 25 people carry the cystic fibrosis gene, which means that in one in 625 couples (25 times 25) both partners are carriers. And as

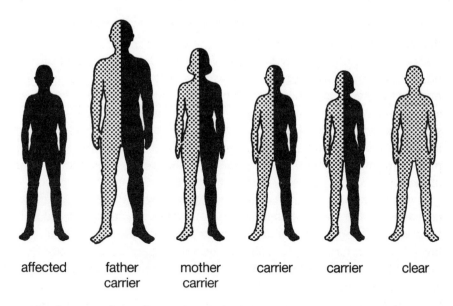

affected   father   mother   carrier   carrier   clear
                carrier   carrier

*The inheritance of cystic fibrosis: the gene for the disease is carried by symptomless
individuals. A child will be affected only if he or she inherits a copy of the cystic fibrosis
gene from both parents. On average, one child in every four born in such a family will
be affected by the disease; two out of four will be symptomless carriers; and one will be
entirely free of the gene. This pattern follows the rules laid down by Gregor Mendel.*

we have seen there is a one-in-four risk that such couples will give birth to
an affected child. The birth rate is therefore 1 in 2,500. On average, those
who are born with cystic fibrosis now live to their twenties, and some to
their forties. The disease spreads a perpetual shadow over their lives, however.

Given the ubiquitous nature of cystic fibrosis in the West, it is not surprising that much effort was put into searching for its cause. But again,
scientists ran up against a solid wall of medical confusion. What possible
common denominator could produce lung infections, diarrhoea, pancreas
blocked with sticky secretions, and sweat that contains high concentrations
of salt? What mutated gene could be responsible and what protein did it
code for in its normal state? For all their efforts, clinicians and biochemists
never found the answer – but the molecular biologists did.

The serious business of hunting the cystic fibrosis gene began in 1981, a
couple of years after Bodmer and Solomon had published their letter, and
just after Davies and Williamson had pinpointed the Duchenne gene. Kay
Davies continued to work on Duchenne, but Professor Williamson, having
demonstrated the power of gene mapping, was thirsty for a new challenge,
and cystic fibrosis fitted the bill. He was not alone. Teams from Toronto,

Cleveland, Berkeley, San Antonio, Salt Lake City and Copenhagen also joined the search for the gene. It proved to be a heady combination.

At the beginning the work was laborious. Without a chromosome to start on, researchers had to go through a wearisome procedure of trying countless numbers of different probes. It was a basic business: test and eliminate, test and eliminate, over and over again, for years. In 1983, the scientists met to swap data, and managed to discard 20 per cent of the genome from their inquiries. Then the molecular biologists went back to their riflips and their probes. In August 1985 they met again, and dismissed another 20 per cent of the genome from further consideration.

It was slow going. At the rate of 10 per cent elimination a year, it could take a decade for the combined resources of half the world's molecular biology laboratories just to find the right chromosome. Then came the breakthrough. Hans Eiberg, from Copenhagen, had been studying blood samples from Denmark, London and Toronto. He was using much older techniques in which he looked at differences between the proteins in people's blood rather than their genes to see if any of their variants were inherited with cystic fibrosis. He discovered that a variant of a blood enzyme called paraoxonase seemed to track with the disease. A marker had been found.

Unfortunately, no one knew on which chromosome the paraoxonase gene lay either, or even if it was made by more than one gene, like haemoglobin! But there was a great deal of data on 'where it was not', says Professor Williamson. 'At that stage, we had a lot of information that suggested that the cystic fibrosis gene was on one of three chromosomes – 7, 8 or 18. When we got together to pool our knowledge about cystic fibrosis and paraoxonase, it became clear that there was one big hole – on chromosome 7.' The hunters were closing in for the kill.

After pinpointing the chromosomal culprit, the molecular biologists went back to their laboratories and ran so many probes over it, they almost ironed it flat! All they needed was one really good riflip that associated with the disease's passage. Within three months they had found five. Teams from Toronto, Salt Lake City and St Mary's reported in *Nature*, in November 1985, that their markers sandwiched a piece of genetic material on the long arm of chromosome 7. Somewhere in the middle lay the cystic fibrosis gene.

By now the race to find the gene was becoming nerve-racking. Gone were the early days of spirited co-operation. There was no hostility, but an air of uneasy recalcitrance now permeated laboratories. There was an obvious reluctance to share information – and credit. As the journal *Science* put it: 'Even within the highly competitive field of human genetics, the search for the cystic fibrosis gene stands out for the intense nature of the rivalry.' Despite these handicaps, however, the different teams continued to edge along chromosome 7, homing in with their probes closer and closer to the gene. Then in spring 1987 Professor Williamson announced he had found 'a strong candidate'. The hunt looked over. It took a further four months for the St Mary's team to show that they had in fact found a different gene, one now called the int-related protein gene, whose purpose remains obscure, and which turns out to be the next-door neighbour to the cystic fibrosis gene. It was eliminated because it was not expressed in gut and lung cells, where the symptoms of cystic fibrosis are initiated. St Mary's had missed by a couple of thousand DNA base pairs – a hair's breadth in genetic terms.

In the end, the honour of the final kill went to Lap-Chee Tsui, of the Hospital for Sick Children in Toronto, working with Francis Collins of Michigan University. In 1989, his team found a lengthy slice of DNA that appeared to track absolutely with the disease. Their observation was quickly confirmed. The gene was expressed in the right place, the gut and lung. Tsui and Collins had found the cystic fibrosis gene.

Today we know that the gene whose mutations cause cystic fibrosis normally codes for a protein that is called the cystic fibrosis transmembrane regulator, or CFTR. Its function is to sit on membranes of cells in the lungs and gut and facilitate the flow of chloride ions. This first clue to its role came from comparing it with other known genes and finding a similarity to one gene involved in the transport of chemicals in and out of cells. In its mutated form, there is disruption to this stream, not just of the ions, but also of water molecules that would ordinarily accompany the chloride. And that is what produces cystic fibrosis symptoms. Mucous secretions in the lung and gut become sticky for the simple reason that not enough water flows out of their cells. Ducts become blocked and infection sets in. The explanation even accounts for the salty sweat. It is not a problem of too much salt, but not enough water. From this one failure emanate all the symptoms of cystic fibrosis. It was the most important medical development in the disease's history. It was also a complete triumph for the business of 'reverse genetics'.

The only trouble was that Tsui's mutation proved to be only one of many. Admittedly, he had found the most common one – a three base pair deletion (codenamed Delta F508) which resulted in the CFTR protein being made without a crucial amino acid, phenylalanine. It is the mutation that accounts for cystic fibrosis's elevated levels among Caucasians and is responsible for about 75 per cent of their cases. But over the next few years, an incredible total of 250 further cystic fibrosis mutations were found to be responsible for the remaining 25 per cent of cases. Most of these are extremely rare. Nevertheless, because it is very expensive and time-consuming to test for more than three or four mutations at one time, all these genetic varieties have limited the efficacy of screening. And that has had one major consequence.

A sizeable fraction of people – about 20 per cent – will be cleared of being carriers, but could still turn out to have the cystic fibrosis gene. 'The bottom line is that we are never going to prevent cystic fibrosis children being born,' says Professor Williamson. 'So we're going to have think very carefully about treatments and cures straightaway.'

On the other hand, the prognosis for cystic fibrosis, both in terms of developing foolproof tests, and ultimately creating treatments, had been improved beyond all measure thanks to the discovery of its genetic roots. Most important of all, the work of those scientists from Toronto, St Mary's and all the other centres has paved the way for the introduction of carrier screening throughout the West. And once these have been launched, mothers and fathers like Sheila and John Rose will be given quite explicit warnings, well in advance, about the dangers facing their unborn children, and foetal screening services offered to them. That in turn raises issues about improving biological education so people can properly appreciate the nature of these warnings, of course. It also means that proper genetic counselling must be made available so carriers picked up by screening can be guided in their decision-making. These issues pose tricky, but not insurmountable, problems – which we shall discuss in the final chapter.

Of course, the three ailments we have concentrated on so far are a mere sample of the progress that has been made in probing the roots of single-gene inherited diseases. Other conditions that have succumbed to the relentless genetic interrogations of researchers armed with their polymorphisms and probes are myotonic dystrophy, a relatively mild muscle wasting unrelated to the X chromosomes; Menke's disease; fragile-X syndrome, a common form of retardation, and many more.

It has been a golden era, one that began almost exactly with the arrival of the new decade in 1980, when Davies and Williamson began their search for Duchenne and Gusella prepared for his hunt for Huntington's. Almost exactly a decade later, that magnificent era closed.

'It was a gloriously exciting time,' says Professor Davies. 'But it was also an enormously laborious one. And the Human Genome Project is changing that. By the mid-1990s, we won't hunt for a gene. We will simply home in on a bit of chromosome, look up all the genes that we know lie there, and select the ones which might have a mutation that fits our symptoms. The top laboratories will not be filled with DNA sequencers, they will be staffed by experts on viruses, cell development and all the other technologies that are needed, not to find a gene, but to put right its mutated output.'

The revolution is not over, of course. It has merely moved on to another phase. And in any case the knowledge already gained has been of inestimable value. It has allowed doctors to tackle a source of suffering that has been passed on through interminable generations. Thanks to one of the most exciting decades in the history of biology, the sins, or more accurately the vicissitudes, of the father, or mother, are no longer being visited upon succeeding generations. The genetic thread that has tied victims to their pasts is being cut. And that refers not just to diseases caused by simple, single mutations, but to more complex causes of illness, such as cancer and immune disorders. These conditions are the subjects of the next two chapters of the Book of Man.

6

Wrong Division

King George VI died in the early morning of 6 February 1952. His valet entered the King's bedroom at 7.30 am, and found that the man who had reigned over Britain through the Second World War had passed away in his sleep. Like millions before and after him, King George left this world a victim of lung cancer. He had been ill for years, and eventually succumbed to complications of an operation intended to correct his condition. George was only fifty-six years old but for most of his life he had been a heavy smoker.

And of course the two facts are connected. We know today that George VI died in middle age because he smoked. Cancer, scientists have discovered, is a genetic condition in which cells spread uncontrollably, and cigarette smoke contains chemicals which stimulate those molecular changes.

In 1952, these facts were only just being uncovered – too late to help George VI; or his brother, the Duke of Windsor (who died of cancer of the throat); or his father, George V; or his grandfather Edward VII, who all suffered from smoking-related illnesses. Like so many other families, the Royal lineage was inured to the notion that the habit was not harmful, an idea that has fatally misled so many for four centuries.

Tobacco was brought to Britain from America in 1586 by Sir Walter Raleigh, and enthusiasm for its addictive pleasures spread quickly, despite the entreaties of the monarch, King James I. He denounced the habit as 'dangerous to the lungs, and in the blacke stinking fume thereof, neerest resembling the horrible stigian smoke of the pit that is bottomlesse', grimly prophetic words that fairly typify the 'wisest fool in Christendom'.

Equally typically, no one treated his ideas seriously, and it was not until the reign of James's distant successor, George VI, that the accuracy of his 'stigian' prediction was realized. Dismayed doctors started to see soaring numbers of lung cancer cases appearing in their surgeries in the 1930s and 1940s. A few guessed that smoking, which had increased enormously then, was involved, though it was not until 1954 that Richard Doll and the late Austin Bradford Hill carried out a conclusive study of more than thirty thousand male doctors which showed that lung cancer was unequivocally linked to smoking.

Even then, the medical profession remained sceptical. Doctors accepted the connection, but argued that it did not prove that the inhalation of chemicals actually caused cancer. This misconception was ironic, for the idea for Doll and Hill's study had originally been put forward by Sir Ernest Kennaway, the British scientist who had discovered that specific chemicals actually cause cancer in animals. In any case, the link between 'chemical abuse' and cancer had already been established in the eighteenth century when the British surgeon, Percival Pott, realized that boys who climbed up chimneys to clean them were liable to develop cancer of the scrotum, the skin bag that contains the testicles. The boys' susceptibilities arose because they were continuously exposed to soot. As Pott put it: 'They are bruised, burned, and almost suffocated, and even when they get to puberty, become peculiarly liable to a most noisome, painful and fatal disease.' Thus Pott not only recognized the first occupational cancer, but also identified a substance – soot – as a probable cause.

However, more than a hundred years were to elapse before his farsighted ideas were confirmed and developed. In 1918 it was found that coaltar, repeatedly applied to rabbits' skin, caused cancers. (The word cancer is Greek for crab, and was probably first used to describe the swollen, crablike veins surrounding a tumour. As the medieval writer Fallopio put it: 'Cancers seize on the surrounding parts with the tenacity of a crab seizing on its prey.') Ernest Kennaway's critical contribution was to identify chemicals, called polycyclic aromatic hydrocarbons, as the most active cancer-inducing ingredients of coaltar. These compounds are also found in cigarette smoke.

The evidence was mounting. Regal lungs, a chimney boy's scrotum, the skin of a rabbit: all were revealing their vulnerability to the carcinogenic predations of certain chemicals, substances which are extremely efficient in damaging DNA, causing the genetic changes that we now know underlie the development of a cancer. And there were intriguing clues emerging from other areas. For example, it was found that cancer, which can occur in virtually any tissue – lung, scrotum, breast, stomach, skin, bowel, bladder, bones, muscle, and blood cells – differs in incidence according to where people live. And these variations also provided important clues about causes, and about the nature of the condition.

Take skin cancer. Rates among white people in subtropical areas, such as Queensland in Australia, have soared this century. Today they are two hundred times higher than those found among dark-skinned people in lands such as India – and that is because white skin provides poor protection against the sun's rays. Strong sunshine is rich in powerful ultraviolet radiation which we now know is another trigger of cancerous genetic changes.

Similarly, it has been discovered that breast cancer in Britain and America is four times more common than it is in Japan. However, Japanese women who settle in the United States subsequently become as vulnerable to the disease as their American counterparts. Somehow they lose crucial protection, and many scientists suspect that diet differences, possibly related to a low fat intake, may be involved.

And then there is the simple question of age. Above all, cancer is a disease of old age. In the West, more than 80 per cent of cases arise in people who are older than fifty-five. The chance of getting cancer under the age of twenty is less than a hundredth that at the age of seventy-five. Even the risk at fifty-five is five times less than that at seventy-five. Clearly, something about the ageing process elevates one' s risks of getting cancer, a phenomenon that has become increasingly obvious in recent years. As infectious diseases, which used to kill us off in early life, have been vanquished in the West, we are living longer, and are succumbing to cancer more often. Today, cancer is responsible for more than a quarter of all deaths in developed nations.

Radiation, chemicals, age – these, then, are some of the triggers that set in motion the grim chain of events that leads to cancer. A full understanding of their handiwork, and an appreciation of the trail of their cellular havoc, has been critical in helping unravel the causes of one of most potent killers to strike at humans.

Today, we know cancer arises as a result of several independent events, each biological accident leading to a change in DNA that combines with the next to produce a tumour. It is the genetic equivalent of an accumulator bet in which several horses from different races are selected, and the winnings from each invested as the stake for the next race. To win, every horse must be first past the post in every race. In other words, several independent events – the outcomes of different races – must occur in succession. Similarly, to contract cancer, at least five or six different biological incidents must take place in succession. And the longer one lives, the more time is available for each event to transpire, and eventually for all to occur together.

One can also compare one's chances of getting cancer with the risk of having a car accident. The longer you drive a car, the more likely it is you will have a crash, although other factors – alcohol, tiredness and road conditions – increase prospects as well. The same is true of cancer. We can raise the odds of succumbing to a genetic accident by exposing ourselves to extra hazards in life, and of these, smoking and baring oneself to strong sunshine without proper protection stand out as especially effective life shorteners.

We know this today. However, it was not until the nineteenth century that the German scientist Rudolph Virchow demonstrated that any cell capable of multiplying can develop into a tumour. In doing so, he opened up the investigation of cancer as a disease of cells.

A tumour (which comes from the Latin word for 'to swell') is a localized abnormal growth and it does not necessarily have the potential to invade or destroy. A surgeon can remove a benign tumour completely, leaving no possibility of further disease. It becomes malignant only when it develops the potential to invade and destroy surrounding tissue. It is then called a cancer. However, a cancer at this stage can still be halted before it spreads too far. Only when metastasis (from the Greek 'change') occurs does a malignant tumour spread to produce secondary growths round the body. Then it cannot be removed simply by surgery or radiotherapy – which explains why it is so important to detect cancers early in their development.

When investigating a cancer a pathologist cuts a piece from the tumour (a process called biopsy) and either freezes it or embeds it in wax so that very thin slices can be cut, stained and placed on microscope slides. And if the tissue, say of the bowel, is healthy, the pathologist will see cells in highly organized patterns – for instance, a series of small sacks or crypts which line the main surface of the bowel. (It looks rather like a long hosepipe with lots of little indentations.) However, if a tumour is present, this regularity is lost, and the more malignant the cancer, the more disorganized the tissue, until eventually there are just lumps of epithelial cells (the type of cell that lines body cavities such as the stomach and the bowel) that bear no relationship to their normally ordered alignment.

So what triggers a cell's sudden abnormal growth? Why does one suddenly make a pell-mell bid to escape its sedately managed existence, take on an independent, uncontrolled vitality and spread wildly through the body? The answer is that changes to a cell's genetic information, stored in its nucleus and which control its ordered continuance, are responsible. And as successive mutations arise (remember our genetic accumulator bet), the developing cancer cell and its progeny acquire increasing powers to divide independently, to multiply more and more rapidly, and to disseminate through the body. The process starts with a single mutation in a cell from which eventually all else follows. This, then, is the basic model of cancer causation that is understood today. It took a long time to be accepted, however.

A key figure in this struggle was Theodor Boveri, a German classics student turned physician who, at the of age of thirty-one, became professor of

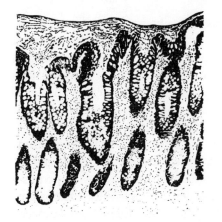 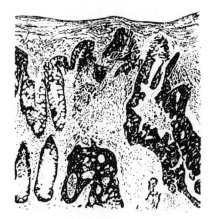

The hallmark of cancer: tissue cells are normally arranged in highly organized ways (left). However, this regularity is lost when a tumour is present (right).

zoology and comparative anatomy at Würzburg University in 1893. He was almost exclusively an experimental scientist (he had a particular penchant for dissecting sea urchins) and published only one theoretical paper, his last. And what a scientific swansong it proved to be – for in it, Boveri proposed that cancers arise from abnormalities that occur in chromosomes during faulty cell division. Boveri's medical background, his interest in cell biology and his understanding of genetics played significant roles in this breakthrough in understanding. The depth of his insight still remains remarkable.

Unfortunately, Boveri's ideas were not published in English until 1929, fourteen years after he died. And even then, available techniques could not substantiate his theory. As a result, Boveri's concept of cancer was not fully accepted until nearly fifty years after his death.

An indication of his farsightedness is provided by James Murray, a distinguished scientist, director of the Imperial Cancer Research Fund, and a pupil of Boveri's. Murray spent years vainly trying to establish that chromosomal abnormalities underpinned the birth of cancers, but eventually had to give up. As he wrote in 1929: 'It is now many years since I relinquished the attempt, after laborious trials, to find constant significant alterations in chromosome numbers, nucleus size, or nucleus protoplasm ratio in cancer cells.' Once again, an idea far ahead of its time had to wait for technology, and the thinking of other scientists, to catch up with it. It could be the leitmotif for the development of genetics.

In the end Boveri was vindicated – in the 1950s when scientists discovered how to prepare and isolate chromosomes so that they could be individually identified. In its first application in cancer research, the technology was used to study chromosomes in leukaemia, the cancer of white blood cells. (The name, given by Virchow, literally means 'white blood'.)

Two researchers, Peter Nowell and David Hungerford, from the Institute for Cancer Research in Philadelphia, decided to look at chromosomes of cells of chronic myeloid leukaemia. To their surprise, they found a strange little chromosome, smaller than any normal type, in all the white blood cells extracted from their leukaemia patients. They called it the Philadelphia chromosome. Theodor Boveri's ideas were finally being vindicated.

It took many more years, however, before the chromosome's strange origins were revealed and its connection with the aetiology of cancer unravelled. In the early 1970s new accurate staining techniques were developed and, using these, Janet Rowley in Chicago showed that the Philadelphia chromosome was actually a combination of part of chromosome 22 and part of chromosome 9. There was a 'translocation' in which the tip of chromosome 9 replaced the tip of chromosome 22 and vice versa.

In other words, the leukaemic cells actually contained two abnormal chromosomes, the Philadelphia and an abnormal number 9 with a longer tip derived from number 22. At some stage during the leukaemia's development a translocation occurs and this gives cells a growth advantage over other white blood cells, clearly implying that there must be some rearrangement of genetic information where the chromosomes break and rejoin, a molecular

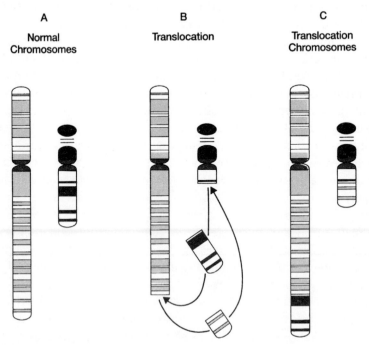

A	B	C
Normal Chromosomes	Translocation	Translocation Chromosomes

A chromosome translocation.

realignment of benefit only to the leukaemic cells. But what exactly is that bonus? What genetic edge accrues in this deadly new configuration, researchers wondered? They found the answers, but from a very different, and seemingly unlikely, strand of scientific endeavour – the study of viruses.

Peyton Rous was working at the Rockefeller Institute in 1909 when a Plymouth Rock hen with a breast tumour was brought to his pathology laboratory. Rous decided to try to transfer the animal's cancer to another hen to propagate it for further study. He minced material from the tumour and injected it into other Rock hens. One developed a tumour and this was used to create tumours in others.

Rous then went a stage further. He broke open his extracted tumour cells, and strained them so finely that not even a bacterium could pass through his filters. Still his tumour derivative induced cancer. The conclusion was inescapable. At least one form of cancer must be caused by a virus, the only known biological entity small enough to pass through the extremely fine filters employed by Rous.

Unfortunately few scientists believed the experiment. One pathologist even told Rous: 'Look here, young man, that can't be a cancer if you found its cause.' Even Virchow refused to accept the idea. Eventually Rous was vindicated, however, and in 1966 he was awarded the Nobel Prize for his discovery – fifty-five years after publishing his original paper, a record deferral even by Nobel standards. Once again our recurring theme of delayed recognition was in operation – for today cancer-causing viruses, or oncogenic viruses as they are also known, have become a major focus of medical research.

The discovery of the Rous sarcoma virus (as it became known) was eventually followed by those of other tumour viruses in rabbits and mice, which opened up the prospect of carrying out controlled laboratory experiments on cell cultures. Normally when tissue cells are cultured, they die out, usually because some unknown intrinsic properties prevent them from dividing to an unlimited extent. Not so with cancer cells. They can be propagated indefinitely to form a permanent strain of cells, or a cell line as it is often called. This, of course, parallels their power to thrive independently in a living animal. And when oncogenic viruses are used to infect normal tissue cultures, a similar phenomenon occurs: colonies of transformed, immortal cells are generated, ones moreover that can induce tumours when inoculated into living animals. Once this was achieved it was possible to ask, realistically: what properties do oncogenic viruses possess which enable them to cause cancers?

The answer came in the 1970s with the discovery of strains of oncogenic viruses which had lost their potential to cause cancer, a crucial lead that allowed Michael Bishop and Harold Varmus of the University of California at San Francisco to pinpoint the extra piece of information carried by oncogenic viruses. Strikingly, this extra segment almost exactly matched a fragment of genetic information present in the virus's host cells' chromosomes, indicating that the virus must have usurped this fragment so it could remain hidden in a cell, and divide along with it. In this way, the virus gained a malign influence that allowed it to divide without limit. These acquired genes were called oncogenes and their discovery earned Bishop and Varmus the Nobel Prize in 1989.

We started looking at oncogenic viruses as a diversion from the story of the Philadelphia chromosome. Now, armed with the knowledge we have gleaned from these tiny tumour factories, we can return to this strange aberrant chromosome. Scientists wondered if it could be that an oncogene lies at precisely the point where chromosomes 22 and 9 break and exchange parts. But to find out they needed to be able to discover if the normal gene which is transformed into an oncogene is located on chromosome 9 or chromosome 22.

The locating of genes on a chromosome is not a straightforward business, and requires a great deal of ingenuity and effort on behalf of scientists. Recently, however, new techniques have been developed to localize sequences on to their chromosomes using cloned pieces of DNA. The sequences can be tagged with a fluorescent dye and then applied to a preparation of human chromosomes in such a way that they will attach only to their matching sequences. Using a fluorescent light and a microscope, you can simply light up the position where the DNA piece has landed and identify its position on a particular chromosome. This, put simply, is the basis of genetic mapping, and it has proved to be fundamental in identifying the causes of diseases, including cancers.

Among the oncogenes that have been found on cancer-causing viruses is one known as abl, an abbreviation derived from a Dr Abelson, who discovered it. And it was this oncogene that was chosen by a team from the Imperial Cancer Research Fund (ICRF) led by Walter Bodmer, one of the co-authors of this book, and scientists at the National Cancer Institute in the United States to create the first genetic map of oncogenes. Their preliminary studies produced a remarkable result. They found the abl oncogene normally lay on chromosome 9, and most probably on the long arm – and that, you will recall, is precisely the translocation point involved in the creation of the Philadelphia chromosome.

'The cloned piece of DNA we were studying was just about a thousand base pairs long, while the length of the piece of chromosome to which we had assigned this sequence was about a hundred million base pairs,' recalls

Walter. 'That would suggest there was only a one-in-a-hundred-thousand chance that the position of the abl gene had anything to do with the Philadelphia chromosome and leukaemia. Nevertheless, the coincidence seemed too good to be true.'

This speculation was soon substantiated by American and Dutch researchers who began an intense, detailed study of the abl gene. They showed that in creating the Philadelphia chromosome, the abl gene was moved from its normal position on chromosome 9 to chromosome 22, where it was juxtaposed to another gene they called bcr (for break cluster region). As a result a totally new gene was established, one that looked like the bcr gene (from chromosome 22) at one end, and the abl gene (from chromosome 9) at the other. This new gene is never found in a normal cell and is unique to the Philadelphia chromosome. It must therefore be one of the key mutations involved in the creation of a cancer, a key 'successful' bet in our genetic accumulator.

The discovery was one of the most dramatic pieces of evidence that genetic changes cause cancer. Since then, other similar discoveries have been made, though such is the pace of development of gene mapping technology, it now takes only a few months to reveal the exact nature of a translocation compared with the nearly quarter of a century of effort that separated the first description of the Philadelphia chromosome and its elucidation at the DNA level.

Of course, this does not explain what an oncogene normally does, nor reveal why, when it is mutated, it helps a normal cell in a step-wise malignancy towards cancer. However, there is an obvious place to look for answers to this question – given that we now know that cancers character-

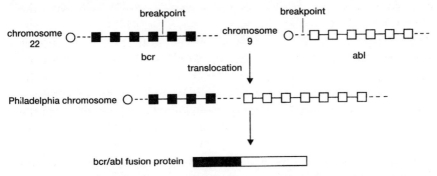

The creation of an oncogene. A translocation occurs in which the tip of chromosome 22 is replaced by the tip of chromosome 9 and vice versa. In the process, an abl gene on chromosome 9 is moved from its normal position and is juxtaposed to another gene on chromosome 22 called bcr. This new gene is known as bcr/abl and its creation is a key mutation in the causation of a cancer.

istically involve cells that grow uncontrollably – and that is the genetic mechanisms which regulate cells' growth and division.

Cells divide only when instructed to do so, otherwise bodily tissues would have no internal organization. These chemical signals are called growth factors, and were discovered at Washington University, St Louis, by Rita Montalcini and Stanley Cohen. Growth factors are produced at independent sites in the body and when one hits a cell, it sets off a trigger telling that cell to divide.

One of Cohen's first discoveries was a growth factor for epithelial cells, the very cells from which many common cancers – breast, lung and bowel – are derived. His growth factor was called EGF for epidermal growth factor, epidermis being the outer layer of skin. (You can catch a glimpse of growth factors in operation when you watch a dog or a cat lick a wound. Animals do this not just to keep the laceration clean, but because their saliva contains a growth factor that stimulates the cells in a wound to divide, so promoting the process of healing.)

The growth factor discovered by Cohen is a protein and it works by attaching itself to another, complementary, protein on a cell surface called an EGF receptor. You can think of the way a growth factor inserts itself into a receptor as being like a key that fits into a car ignition. Once it is in, it turns on an engine – one that drives the motor of cell division. You can therefore see why a cell must not produce its own growth factor. If it did, it would simply roar off, stimulating itself to divide continuously and incestuously – like a cancer.

Growth factors and their receptors are therefore of fundamental importance, as was realized by Mike Waterfield, another ICRF researcher, who decided to investigate the detailed amino acid structure of the epidermal growth factor receptor. One day in 1984, he typed details of one stretch of these amino acids into his computer and instructed it to search for matching sequences in a database of all DNA and protein sequences. (As in so many other branches of science, the use of computers has come to dominate research in genetics and molecular biology.) Minutes later, it flashed up a result that was to send him scurrying through the ICRF building: his growth factor receptor sequence almost exactly matched a sequence from an oncogene called erb-B.

Waterfield was excited for good reason. His discovery showed that in its normal state the erb-B oncogene is an epidermal growth factor receptor. He was particularly excited because a few months earlier he had similarly matched another oncogene and a growth factor, called PDGF. (PDGF stands for platelet derived growth factor, a platelet being a small cellular component of blood involved in wound repair and clot formation.)

In combination, these results provided the first clues to the origins of oncogenes, of which, we now know, mutated growth factor receptors are a common variety. In the case of the EGF receptor oncogene, it turned out to be an abnormal form of the epidermal growth factor receptor and was telling the cell to divide over and over again without waiting for proper 'outside' instructions. In the other oncogene, the platelet growth factor was being produced by the cancer cell, again creating a deadly self-stimulation to divide.

The list of known oncogenes has grown substantially since 1984 and all its constituents have been connected, somehow, with the control of cell growth. Between inserting a growth factor key into a receptor lock, and turning the ignition of the cell division engine, many complicated procedures must take place. We know now that any one can be disturbed by mutations, pushing a cell away from normal, regulated division down the road to deadly, uncontrolled expansion.

Exposing this vision of unbridled self-replication has been one of the great achievements of modern medical science. Firstly, by studying viruses, scientists discovered that pieces of DNA called oncogenes triggered cancers and that these genetic entities bore striking similarities to normal DNA sequences. Subsequent research showed that the progenitors of these oncogenes (which are sometimes called proto-oncogenes) were genes which normally direct the growth and division of cells in the body. In their mutated form they cause those cells to proliferate without limit – so producing cancer.

However, we should not forget our earlier description of cancer as a biological wager, an accumulator bet in which many outcomes must combine to produce final victory for a tumour. That uncontrolled running of a cell's engine, the mutated revving of a cell growth receptor that we have just discussed, represents only one of these outcomes. There are others, and some are involved in the very opposite process to cell division – in other words, they are concerned with the suppression of cell growth. Intriguingly, the discovery of this parallel strand in the unravelling of the molecular biological roots of cancer comes from a far more traditional approach to the subject of genetics: the study of family trees. This may seem as strange as studying viruses as cancer causatives, acclimatized as we are to the notion that such diseases have mainly physical and chemical causes that emanate from our environment. Nevertheless, cancer can also be inherited, or at least a predisposition to it can be. And investigation of such cases, although generally rare, has provided powerful new clues to the causes of cancer.

Take the recent case of Linda Young who developed a growth in her right eye when aged only eight months. Retinoblastoma, a childhood cancer of the retina, was diagnosed and so the eye was removed before the tumour could spread to her brain and kill her. The disease often runs in families, and when it does there is a strong chance it will occur in both eyes. Linda's other eye seemed normal, however. Then Linda's brother John was born, and by the time he was seven months old, he too had developed a retinoblastoma. So Linda was again examined, now aged eighteen months, and this time a cancer was found in her remaining eye. Too extensive for treatment, it was also removed.

Poor Linda's story is particularly poignant because her blindness might have been prevented if more frequent examinations of her left eye had been carried out. These might have picked up the retinoblastoma which could have been treated with X-rays, so saving her sight. However, Linda's case was complicated because neither of her parents had retinoblastoma. As a result, when it was first diagnosed in Linda, she was not considered to be a 'familial' case. Only when her brother John's condition was discovered was the family susceptibility revealed and the risks to both Linda's eyes recognized.

Retinoblastoma is fortunately quite rare, affecting only one in 100,000 children of whom more than half inherit it from a parent as an apparently simple Mendelian dominant trait, though to be precise they do not inherit a cancer itself, but a predisposition to get one. The remaining body of cases arise sporadically and without warning in a family. And it was by comparing these with familial examples that Al Knudson, working at a cancer centre in Philadelphia in 1972, developed an ingenious hypothesis that not only explained the two forms of the disease, but also shed light on many other cancers.

Knudson realized that if a genetic change is a key event in making a cell cancerous, that metamorphosis could occasionally occur, not in a body cell, but in a germ cell – in a sperm or an egg. This alteration would then be passed on to every cell in the resulting embryo. These individuals, Knudson argued, had already climbed the first rung on the ladder towards cancer. Even before birth, all their cells had notched up that first successful wager in our biological accumulator bet.

And that critical step, Knudson argued, involves the complete knocking out of the function of a gene that must somehow play a crucial role in preventing or 'suppressing' a cell from becoming cancerous. Its function is therefore the opposite of a dominant oncogene, whose role is to promote cell growth and division. A suppressor gene is supposed to shut it down.

However, a lack of one suppressor gene in a cell is not fatal on its own. We acquire our genes in pairs, as we know, and if one does not do its assigned task, then its partner will generally carry out its missing function. It is a lonely responsibility, however, and such biological exposure is clearly revealed in retinoblastoma. Eventually, a mutation wipes out the partner gene in one of the thousands of retina cells in the eye, and without any suppressor genes to call on it is exposed to the process of uncontrolled division. The result is unrestrained growth, and retinoblastoma. With spontaneous cases, where there are no inherited predispositions, two mutations must occur, and the chances that both will arise naturally in the same cell, or line of cells, is, of course, very low.

But where was the suppressor gene involved in retinoblastoma? Where did this cell regulator sit on our chromosomes and how did it operate, scientists wanted to know. The answer came, as it has with so many other gene 'breakthroughs', from spotting a chromosomal abnormality. A few retinoblastoma cases were found to have a small deletion on chromosome 13, and that is where the suppressor gene must lie, researchers realized. Subsequent linkage studies placed it next to a gene for an enzyme called esterase D. The two were so close that they were often deleted together in retinoblastoma patients.

It was a critical discovery, for scientists found that the esterase D enzyme acted as a marker for the retinoblastoma suppressor gene, exactly mimicking the behaviour that Knudson had predicted. Firstly people inherited a knocked-out version of a gene, leaving just one normal functioning esterase D gene behind, and then there was a second event which removed this remaining activity.

By now researchers were so close to the retinoblastoma suppressor gene that it only took a measure of good luck and neat laboratory management to find it, and clone it. As a result, it is now possible to determine if a retinoblastoma case is due to a mutation inherited from a parent, or if it is sporadic. Sadly, science had not properly equipped itself at the time when Linda Young's first retinoblastoma was found. In future, now that we have uncovered the guilty gene, or more precisely the lack of it, eye cancers like Linda's should be caught early enough to carry out effective treatment.

The role of the retinoblastoma gene is clearly preventative, since it is only when its function is eliminated that cancer develops. That is why it is called a 'tumour suppressor' gene, and its function contrasts markedly with the oncogenes that were described earlier in this chapter. Their presence, not absence, does the damage in a cell. Indeed, scientists suspect a tumour suppressor gene may actually stop a growth factor from being made in inappropriate circumstances.

In any case, the crucial point is that cancer is tied up with the control of growth. Think of our car engine analogy. We can view a mutated oncogene as the genetic equivalent of having the accelerator pedal stuck to the floor of a car. By contrast, a malfunctioning suppressor gene is the cellular counterpart of losing your brakes. Either way, you go careering off into danger.

Now retinoblastoma is a very rare cancer. Only twenty to thirty new cases occur each year in the United Kingdom, figures which compare with the nation's annual catalogue of 25,000 new cases of bowel cancers. We may therefore think that the unravelling of its secrets is of only limited medical value. In fact, the critical point about Al Knudson's brilliant analysis is that it also applies to common cancers. Far from being a bit player in an obscure medical drama, the mutated suppressor gene can be viewed as a genetic actor on a very large stage.

Let us consider the example of Tom Hill, who began to suffer severe constipation and abdominal pains, complaints that were eventually traced to a bowel obstruction that was found to be caused by a cancer of the colon. Tom was only fifty-three, the same age his mother had been when she had died of the same disease. Tom was operated on, and his surgeon found that his colon was carpeted with a series of minute bumps called polyps, precancerous growths from which bowel cancers often arise. And crucially, polyps are often inherited. Such a condition is called familial polyposis, and unless victims are treated, by having their colons removed, they are bound, eventually, to get cancer. Sadly, Tom's cancer had already spread too far, and although the primary cancer was removed, he died from secondary growths three years later.

A precursor of cancer: a polyp, a growth in the bowel from which cancers often arise.

When the familial nature of his disease was explained to his relatives, including his brother Robert and his children, they agreed to be screened for polyps, a relatively straightforward procedure called sigmoidoscopy or colonoscopy which involves pushing a tube with a lens and a light at the end up a person's rectum. A doctor can then see whether the colon is lined with polyps. And Robert had hundreds of them. Clearly, he carried the polyposis gene, though fortunately none had yet become cancerous. His colon was cut out, effectively removing any danger that he might succumb to the disease that claimed his brother.

Polyposis is also quite rare in comparison with other cancers (only in 1 in 5,000 to 1 in 10,000 individuals are affected), and bowel cancers linked to it account for only about 0.5 per cent of all cases. But once Al Knudson's ideas about retinoblastoma had been confirmed, scientists realized the gene involved in polyposis could have a far more wide-ranging impact. But first they had to find out where it lay. And once again, a visible chromosomal abnormality provided the critical clue.

The case in question involved Hugh Dobson, a forty-two-year-old, mentally retarded, life-time inmate of a developmental centre in the United States. Hugh had been operated on for bowel cancer, which is unusual for someone so young. Then, in February 1984, doctors realized he was probably suffering from secondary growths, and so he was operated on again. Not only did surgeons find cancers in his rectum and colon, they found more than a hundred polyps. It was as if he had familial polyposis. Yet, neither of his parents had the condition.

A few weeks later, Hugh died of complications, leaving doctors baffled about his strange combination of ailments and abnormalities. So Lemuel Herrera, his physician, and Avery Sandberg, a geneticist, decided to carry out a 'genetic autopsy' to study Hugh's chromosomes – and found a deletion right in the middle of the long arm of one of Hugh's chromosomes 5. The mystery was resolved, for the deletion must have knocked out the polyposis gene, as well as other genes nearby. This explained why Hugh developed cancer so young, and also accounted for his other abnormal conditions. In short, the polyposis gene must lie on chromosome 5 and must act as a tumour suppressor gene like the gene on chromosome 13 that normally helps prevent retinoblastoma. It was the lead researchers had been waiting for. Within a year, a group led by Walter Bodmer had found a DNA marker closely linked to polyposis in a number of families, one that lay exactly where the deletion had been found in Hugh's cells. If nothing else, his case illustrates starkly the importance of how much vital general information can be gleaned from a seemingly obscure, particular and abnormal individual. Medical science is predatory in the way it will seize on and extract maximum results from a single, illustrative case.

Once the polyposis gene had been localized, Bodmer's team found that at least 40 per cent of bowel cancer cases involve alterations to chromosome 5. However, it took another five years before American and Japanese teams cloned the polyposis gene in 1991. It is now clear, after its mutations have been cloned, that they really do lead to the knocking out of gene function, as predicted.

And now the gene has been identified, the children of people like Tom and Robert Hill can be tested to see if they are at risk of developing polyposis, even before polyps have developed and without using intrusive devices such as a colonoscope. Nor does our story stop here; many other tumour suppressor genes are now being sought. One in particular, a gene that lies on the long arm of chromosome 17, is explicitly involved in cancers of both the breast and the ovaries.

In addition, there is a gene on the short arm of chromosome 17 that codes for a protein called p53 (a term that merely describes its size) that is usually produced in very low amounts by cells, and therefore attracted little scientific attention. Then it was discovered that many cancer cells produce enhanced amounts of p53, and scientific interest was re-kindled. However, p53's true role in human cancers was not properly appreciated until visible abnormalities were discovered on chromosome 17, deletions associated with bowel cancers, and which were found to lie exactly where the gene for the p53 protein was known to lie. So scientists began to screen patients with bowel, and subsequently other, cancers for p53 abnormalities. To their considerable surprise, they found that at least half of all human cancers have p53 alterations which increase levels of the protein. These changes appear to be very similar to tumour suppressor mutations. In other words, loss of normal p53 function leaves humans vulnerable to cancer. Indeed it has even been found that a few rare families have inherited p53 mutations which give rise to susceptibility to different cancers, including breast and bone.

So we can see from the scientific detective work outlined in this chapter that we know there are at least two basic types of genetic alteration which can lead to cancer: oncogenes which actively encourage cells to divide at inappropriate moments, and suppressor genes which sometimes fail to stop this unwanted multiplication. However, we should not think that these two alterations wreak their destructive effects in completely separate ways. Often they will operate in tandem. Indeed, that is probably the usual route by which a cancer develops. For instance, there will be a first stage in which a mutation on one chromosome wipes out the function of a suppressor gene. A line of cells that lack its crucial protective prowess will then develop, such as those which form polyps in the colon. Then an oncogene may be activated in that cell line, causing some of these cells to expand. Then further losses of suppressor genes on other chromosomes will lead to further expansion and wild and more irregular growth until an aggressive tumour is produced. That, then, is how our biological accumulator bet works. These are the accumulated 'winnings' that lead to cancer. And we can also see clearly from this description why cancer is a disease of old age. Each of these mutations will occur at a given rate, and for all of them to take place in a particular carcinogenic sequence usually requires the passage of many years. Equally, from this summary, we can see why many scientists view old age as a simple process by which we accrete more and more harmful DNA mutations.

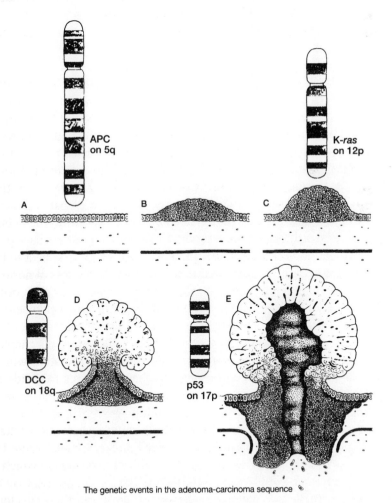

The genetic events in the adenoma-carcinoma sequence

How various mutations in genes (APC, K-ras, DCC and p53) accumulate to cause a cancer.

No doubt, if they were still alive, scientists such as Theodor Boveri and James Murray would be heartened, though perhaps a little surprised, to see how right they had been in their ideas about cancer. What they could not have foreseen were the extraordinary improvements in technology and understanding of genetics and cell behaviour that have been needed to establish their ideas. Today we face the challenge of how to use this new information to tackle cancer effectively.

Of course, prevention is the first priority, and top of the list comes cigarette smoking. We not only know the exact nature of many of the chemical components of smoke, but we have also learned what mutations they produce in smokers' genes, including p53. And the picture is stark: cigarette smoke causes the vast majority of lung cancer cases, and a great deal of other ill health, and death. Indulge in the 'blacke stinking fume' at your peril.

Second on our list is diet. As we pointed out earlier in the chapter, the food we consume is implicated in several types of cancer, as is illustrated by the extraordinary differences that are found in the incidence in breast cancer between Western and Japanese women. However, we don't know exactly what aspects of diet are important. Perhaps it is fat, maybe it is something more subtle. In addition, food is incriminated in bowel cancer, and scientists have variously claimed that fibre, fresh fruit and vegetables reduce risk while red meat raises it. In reality, we do not yet know if these claims are true and it may only be through the basic study of genes, such as the polyposis gene, that we will gain an answer. One particularly encouraging avenue has been opened up by researchers who have created a precise mouse model for polyposis. In other words, they have reproduced the human ailment in rodents by finding a mutation in the equivalent gene. That means that by manipulating these animals' diets, they can find the best way to reduce incidence of polyps in mice, highlighting in the process the way to achieve the same feat in humans, a vital development in the battle to prevent bowel cancer.

Then there are viruses. Despite considerable research, few viruses have been linked to the major human cancers. However, they are implicated in a few, moderately widespread varieties, such as Burkitt's lymphoma, a cancer of the jaw that appears in clusters in some areas of Africa. A virus, discovered by Anthony Epstein and Yvonne Barr, has been found in nearly all Burkitt's lymphoma cases. However, outside Africa, the disease is virtually unknown, which suggests the virus is a necessary, but not sufficient cause of this cancer and that other local factors must be involved. In addition, liver cancer is associated with hepatitis, a viral disease; and cervical cancer is linked to certain strains of human papilloma virus (which is distantly related to the virus that causes ordinary warts), although just as with Burkitt's lymphoma, many people can carry a cancer-causing strain without developing tumours.

Nevertheless in these latter two examples, liver and cervical cancer, vaccinations are now being considered as valid approaches to their prevention.

Radiation is another frequently quoted cause of cancer, particularly leukaemias, and especially with regard to some fear about nuclear power installations. And certainly radiation causes cancer by triggering genetic changes. Nevertheless, the radiation levels to which we are normally exposed are simply too low to make a major contribution to cancer incidence. Obviously we should take care to avoid unnecessary exposure, though the cost involved should be properly balanced against the risks, which are often minimal.

And the same goes for chemicals. We know carcinogens are found in coal-tar and cigarette smoke, though the list of known cancer-causing chemicals is obviously much greater than this. For instance, polyvinyl chloride has been clearly identified as a cause of bladder cancer in workers who use it in factories. Indeed calculations suggest that chemicals in the workplace account for around 5 per cent of all cancers. This is not a huge fraction, but it is a substantial one, and recently companies have begun to test employees to determine if any are genetically susceptible to cancers that might be triggered by chemicals used in their workplace. As we shall see in subsequent chapters, this screening process – although clearly beneficial to susceptible individuals – is not without controversy.

Avoiding risk is obviously crucial to cancer prevention, but it is not always practical. We cannot avoid ageing, for example, much as we would like to. Early detection of tumours is therefore of critical value as well. If, for example, a bowel cancer is detected in time, then a simple, surgical removal effects a complete cure. That is why the extraction of polyps, which may turn cancerous, is so important. Removal of the polyp eradicates the risk of bowel cancer.

But screening a population using colonoscopes, even occasionally and even only those over fifty years old, is cumbersome and expensive. To be effective, less invasive methods of early detection are needed. And one intriguing possibility may be to seek cancerous mutations amongst cells that are shed from a cancer, or even a precancerous growth, at an early stage. Take bowel cancer. Large numbers of cells are constantly sloughed from the bowel's lining into the stool. If there is a precancerous growth or an early cancer there, then its cells will also be discarded. And of course this biological detritus will contain the telltale mutations indicative of an early cancer, though detecting it will not necessarily be an easy process. At a cancer's early stage, one of its cast-off cells might only be present against a background of a million normal cells. Fortunately PCR, or gene amplification, which we first encountered in Chapter Four, is perfectly suited to this delicate but vital task. It can create multiple copies of a stretch of oncogenic DNA and so pinpoint the deadly growth that is beginning to swell within the bowels of a patient. And it is exactly this approach that is now being tested by a number of laboratories, including Bodmer's.

Finally there is the question of curing cancer. The extraordinary wealth of knowledge about genetic changes in cancers has dramatically raised prospects for new, effective treatments and cure. Can we find simple chemicals, for example, that will block the promoting effects of growth factors that are produced by cancers working incestuously on themselves? Or can we find ways to interfere with the function of a mutated p53 gene so that the brakes can be restored to a cell's division process?

It may even be possible to use the body's own defences to recognize the genetic changes that take place in cancer cells. The idea that the body's immune system, which recognizes foreignness in viral, bacterial, parasite and fungal infections, might also recognize an element of foreignness in a cancer cell, is an old one. The trouble has been that cancer cells, as this chapter has explained, are very similar indeed to normal bodily cells, with only slight genetic changes at their hearts separating the two. That is why cancer cells can fool the body's normal anti-disease defences so effectively. However, recent new understanding of the body's immune system has given this approach a new lease of life. But to see how that might operate is a topic for the next chapter.

7

Hunter Killers

Throughout history, mankind has been prey to illnesses that have swept over us in plagues, epidemics and bursts of contagion that have wiped out millions of lives. Some of these diseases have appeared suddenly, like Aids. Others seem to have been with us for ever, such as smallpox, one of the most notorious and pernicious ailments to strike at humans. Its grim effects have been chronicled as far back as 1122BC, when the Chinese recorded a series of cases of 'the spotted plague'. And we can trace its effects back even further, for instance to the days of the pharaohs, such as Rameses V whose mummified head bears the scars of an attack. Even in modern times smallpox has decimated populations. In 1967 an estimated two million people died of the disease, making it as widespread a killer as malaria. Yet only a decade later, smallpox – characterized by fever and the appearance of pustules and often fatal toxic rashes – was wiped off the face of the Earth by a World Health Organization programme of vaccination through which millions of people had their immune systems stimulated in anticipation of an attack by the smallpox virus. It was a stunning achievement, and for the necessary technology the world can thank one man, the eighteenth-century English country doctor, Edward Jenner.

The ninth child of the Reverend Stephen Jenner, Edward was born on 17 May 1749 in Berkeley, England. While a child, a smallpox scare in Jenner's village led to his being inoculated with pus from a victim, a process called variolation. It was thought better to induce a mild smallpox infection than face a full, possibly fatal, attack. It was not an innocuous business, however. Jenner first had to undergo 'bleeding until the blood was thin, purging until the body was wasted and starving on a vegetable diet to keep it so', as he later recalled. It was to no avail, unsurprisingly. Despite these drastic

Edward Jenner.

preparations, Jenner suffered a severe attack of smallpox from the variolation that left him frail for the rest of his life, and very mindful of the disease's debilitating effects.

Then, as an apprentice surgeon, Jenner came across a rather assured young dairymaid who had come to him with a skin rash. 'It cannot be smallpox, because I have had the cowpox,' she told him calmly. 'No one who has had cowpox ever gets smallpox.' Now cowpox was a fairly uncommon disease even in those days, and its victims were generally farm workers who picked it up from infected cows' udders and teats. It was also a fairly harmless condition, its most notable symptoms being inflamed spots or pustules about the joints and tips of fingers.

The dairymaid's words stuck in Jenner's mind, and for years he made notes about the fate of cowpox patients, many of whom appeared to escape the ravages of smallpox. Eventually, thirty years later, Jenner, by then a Fellow of the Royal Society, decided to put his observations to use. First he selected a donor, a young woman called Sarah Helmes who had contracted cowpox through a thorn scratch on her hand while milking cows belonging to her father, a prosperous Gloucestershire farmer. Then he picked a vaccinee, eight-year-old James Phipps, the son of a labourer who often worked for Jenner. On 14 May 1796 the lad held out his arm so that Jenner could make two incisions, 'each about half an inch long'. Into these cuts Jenner dipped some of the liquid that had seeped from a pustule on Sarah Helmes's hand. The world's first vaccination had been effected.

Six weeks later, on 1 July, Jenner inoculated the lad with smallpox pus. 'Listen to the delightful part of my story,' wrote Jenner to a friend. 'The boy has since been inoculated for the small pox which, as I ventured to predict, produced no effect. I shall now pursue my experiments with redoubled ardour.' It was certainly a momentous event. Jenner had managed to protect James Phipps against one of the most virulent ailments known to man. Yet his achievement was not without its irony. Without doubt, Jenner placed the life of young James Phipps at risk, for he had only circumstantial evidence that his trial would work. Modern ethical committees which approve medical trials would have no doubts about their reaction to such an experiment. They would have banned it outright.

What Jenner had realized was that cowpox must be a similar disease to smallpox, but much milder in its effects. It could therefore provide protection without risk. As he wrote: 'It seems as if a chain that endures throughout life has been produced in the action or disposition to action in the vessels of the skin.' Jenner not only laid the foundations for eradicating smallpox, but also established the basis for our understanding of how the body protects itself against infection. He even bequeathed us the word 'vaccination', which

Contemporary cartoon: the terrible effects of the inoculation.

literally means 'encowing'. (In recognition of Jenner's achievements, the British parliament, after long debate, decided to award the great scientist a sum of £10,000, a considerable amount in 1802, though the gesture was somewhat spoiled by the Treasury which first delayed the payment for nearly two years and then deducted over £900 in tax!)

Since the development of Jenner's smallpox vaccine, a host of other, formerly lethal ailments have succumbed to the power of inoculation. Polio, which causes infantile paralysis, has almost disappeared from Western nations thanks to vaccines that can be taken just by eating a lump of sugar dipped into a weakened or attenuated strain of polio virus. Similarly, measles, whooping cough and diphtheria have also nearly been vanquished. But how? How does the body remember a previous exposure to viruses or bacteria so that its defence mechanisms are prepared and alerted? What is the nature of the defence mechanism and how is the invading organism killed or neutralized? These are important questions to say the least, for they raise issues that go to the very heart of the mechanisms that keep us alive and which protect us from the constant biological predations of infectious organisms.

The first answers were provided through the study of bacteria (discovered by Pasteur in 1862) and in particular the bacterium that causes diphtheria. By filtering diphtheria bacteria, a toxin was isolated, a chemical poison that, when injected into a body, would cause lesions and which could kill an animal on its own. In other words, the diphtheria bacterium was found to be harmful because it carried a toxic cargo, like a minuscule bomber that would transport and drop its chemical bombs through the body. The crucial discovery was the finding that serum from animals which survived this toxic 'attack' provided protection against diphtheria when injected into other animals. Somehow these animal survivors produced an anti-toxin which specifically neutralized the diphtheria toxin and no other microbiological agent.

These anti-toxins are known as antibodies and they are generated when a foreign substance 'invades' an animal. Antibodies specifically bind to the chemical features of that foreign material, each feature generating its own unique set of antibodies. (Substances that elicit the production of antibodies are called antigens.) This capacity to recognize specific foreign chemicals gives antibodies an extremely powerful role in the body's defence against invading micro-organisms. But antibodies can also pose difficulties.

Antibodies are the reagents which were used to define the ABO blood groups as we described in Chapter Two, and are the cause of transfusion problems with unmatched blood. If incoming blood is of the wrong type, say type A, and the recipient is type O, then the latter's antibodies will attack cells in the transfused blood causing severe, possibly fatal, side-effects. The problem lies with the clumping of blood, which Landsteiner revealed in his classic experiments and which he used as his basis for categorizing people as being A, B or O blood types. This occurs, we know now, because antibodies latch on to antigen sites on the foreign red cells and so hold cells together, an antibody being able to cling to two separate cells. These cells get linked together in clumps, and as a result they can cause dangerous clogging of arteries. And it was the recognition that foreign blood, even if it came from the same species, was treated by the immune system as if it were an invading organism that paved the way for modern blood transfusion services.

This concept of foreignness is an intrinsic, genetically determined one. The body is declaring its individuality and it can do so in many different ways, not just through ABO blood typing. Some of these are trivial; some are not, as we can see from the case of Mary James who lost her second baby after having had a perfectly normal first pregnancy. Weak from loss of blood, her surgeons arranged for a transfusion. She and her husband were both type O, so it seemed appropriate to use his blood. To the hospital's dismay, she reacted violently to it.

One of Mary's doctors was Philip Levine, who had worked with Landsteiner (the discoverer of ABO and other blood types whom we met in Chapter Three), and he realized that Mary's reaction might be due to an unusual antibody reaction. So he tested her serum against her husband's red cells and found that they clumped. Mary's blood had set up an antibody reaction to her husband's blood and, more to the point, must have done so during her pregnancy in reaction to her foetus's paternally inherited blood type. Her unborn baby carried a factor missing in Mary that triggered an antibody attack from her blood and which caused a severe form of haemolytic anaemia in the foetus, a condition in which its red blood cells were broken open and destroyed. Mary's first pregnancy had survived only because her immune response had not been given enough time to build up its reaction against her baby.

This single case, reported by Levine and Stetson in 1939, was a landmark. For here was an example of a disease caused by the immune system that had nothing to do with infection but was due to naturally occurring differences between husband and wife. This disease, now called 'haemolytic disease of the newborn', is associated with rhesus types positive (Mary's husband) and negative blood (Mary herself), the name rhesus coming from Landsteiner who had detected a similar pattern of reactions in experiments with rhesus monkey blood. We now know that rhesus positive blood and rhesus negative blood are created according to basic Mendelian rules. Negative blood is produced by an individual who has two copies of a gene which cannot produce the rhesus positive antigen. Positive individuals have either one or two copies of the gene that makes the rhesus positive antigen. Such genes operate as a separate category of blood markers to the ABO system so that people can have blood that is O positive, A negative, or whatever.

Blood transfusions are really only special forms of tissue transplantation. We can transplant kidneys and hearts today and we can also carry out bone marrow grafts, but they too have to be very carefully matched. Only those exchanged between identical twins survive without problems, because the body recognizes inherited differences that are present on most people's tissue. The immune system will therefore reject an incoming graft just as it does an infection. These differences are much more extensive than those involved in blood typing, which explains why transfusions are relatively easy to carry out compared with transplants.

These problems were only recognized early this century and their analysis only began thirty years ago, though there has been an explosive growth in research with the development of recombinant DNA technology and modern methods of genome analysis. The crucial discoveries began with study of animal cancers. When researchers transplanted a tumour from one animal to another, they found it did not grow, not because of the nature of the cancer, or because of individual susceptibility, but because the tumour was being rejected as a foreign tissue, just as any normal tissue graft would be. It was Clarence Little at the Bar Harbor Laboratories in Maine who realized the problem and developed inbred strains of mice within which tissue transplants could be carried out because the mice no longer differed enough to recognize each other's tissue as foreign.

However the real motivation to tackle the problem of transplants did not come until the Second World War, when effective ways to treat burns led Peter Medawar to the correct interpretation that graft rejections are immune responses. Medawar was working at Glasgow Royal Infirmary where he teamed up with a surgeon, Tom Gibson. They decided to carry out two different types of transplants: homografts, in which skin from one person is grafted to another, and autografts, in which skin from one part of a person's body is moved to another site. Their first patient was a Mrs McKillop, an epileptic who had burned herself badly after falling on her gas fire.

The pair grafted small pieces of a donor's skin to her and also made some autografts. The two sets of transplants initially behaved the same way. But after a few days, the donor's homograft began to show signs of invasion from Mrs McKillop's white blood cells. Then a second homograft was carried out using the same donor. It was immediately set upon by white cells and destroyed. The autografts all healed normally, however.

As Medawar put it: 'The skin homografts were rejected by an immunologic process . . . by the same general kind of specific adaptive response which daily leads to the elimination of bacteria or viruses or other organisms foreign to the body.' And that response must lie with basic genetic incompatibilities between host and donor. Landsteiner suggested that blood group differences might account for those reactions, and this idea was followed by Peter Gorer at University College, London. He showed that there were such blood group differences in mice. In doing so he discovered a mouse blood group system called H2. For mouse transplants to survive, donor and recipient had to be H2 matched. For a successful transplant the donor must not possess antigens that are absent from the recipient, in other words.

Unfortunately, the exact mouse analogy did not hold for humans – because the crucial immunological difference did not reside with human red cell blood groups. The clues to the solution of this problem came from Medawar. He injected red blood cells from skin donor rabbits into recipients before carrying out skin grafts. This had no effect on the graft's survival. In this case, the rabbits behaved immunologically like humans. But when he injected white cells, this led to an acceleration of graft rejection. It was as if the recipient rabbit had been vaccinated against the transplant, implying that white cells share some determinants with the skin tissue, and therefore hold the key to the success of transplants.

But how could these different white cell types be revealed? The answer was provided, in the late 1950s, when researchers realized that pregnant mothers, in general, might make antibodies to their foetuses' white cells just as rhesus negative mothers make antibodies to rhesus positive offspring. So they started testing serum from women who had recently given birth to see if antibodies were present, ones that would clump with white cells from their husbands, and found that about a quarter of women who have had two or more pregnancies produce antibodies that react with their husbands' white cells, although they do not cause clinical problems, in contrast to their rhesus factor equivalents. Here then was a source of antibodies that could be used to define white cell types to see if they could be matched for transplantation, just as red cell groups can be matched for blood transfusion. The eventual result was the definition of a highly complex pattern of differences between individuals which we now call the human leukocyte antigen (HLA) system.

The HLA system consists of biological markers that cover cells and act as biochemical signatures. These HLA markers are classified into six different types, labelled A, B, C, DR, DQ and DP. Each of the first three – A, B and C – is coded for by a particular gene, while the latter three – DR, DQ and DP – each have more than one subgroup, and are coded for by more than one gene. A total of ten genes code for these HLA system types. And as our chromosomes come in pairs, each of us could therefore have as many as twenty different HLA types. There are more than two hundred variants of these ten genes, and these are numbered in sequence. Thus a person's HLA type might be Al, B8, C3, etc. Each set of HLA types differentiates one person's cells from those belonging to another, and their variation lies at the heart of transplant failure or success.

The genes responsible for at least two hundred individual HLA types have now been cloned and sequenced, revealing a highly complex region of the human genome on the short arm of human chromosome 6. But the discovery of the system depended on simple experiments of mixing serum from one individual with white cells of another, noting the reactions, and then doing a series of statistical analyses to interpret the results. There are literally billions of combinations of the two hundred or more individual HLA types, so that finding two unrelated individuals whose HLA types are the same is the genetic equivalent of looking for a needle in a haystack. A person's immune characteristics are therefore as exact as a fingerprint, a feature that has been exploited, as we shall see in Chapter Ten, to identify a person from their DNA.

But if a person's HLA type is so specific to him or her, how is it ever possible to give them a kidney or heart transplant? The answer is simple. The genes that code for the HLA system occur in a tight cluster that is nearly always inherited as a block. This means that within a family of brothers and sisters there are usually only four different combinations of HLA types, namely, those that arise from the four possible combinations of pairs (one

	DP		DQ		DR					
GENES	DPB	DPA	DQB	DQA	DRB1	DRB2	DRB3	B	C	A
	1	1	1	1	1	1	1	1	1	1
NO. OF VARIETIES	2	2	2	2	2	2	2	2	2	2
	3	3	3	3	3	3	3	3	3	3
	4		4		4			4	4	4
			5		5			5		5
			6					6		
								7		

The body's immune signature. A total of at least ten genes code for our HLA system, one each for the A B C markers, and two each for the DP and DQ markers and three for DR. In turn, each gene comes in many different varieties, so that a person's immune characteristics are almost as exact as a fingerprint.

from each chromosome 6) that come from each parent. If we label the father's two groups of HLA genes as W and X, and the mother's as Y and Z, we can see each child has an equal chance of inheriting four different combinations: W and Y, W and Z, X and Y, and X and Z. There is therefore a one in four chance that someone's brother or sister will have exactly the same HLA typing as theirs. That is why kidney and heart transplants, and especially bone marrow grafts, often use donor brothers or sisters.

However, not everyone who needs a transplant is fortunate enough to have an HLA-matched sibling. Families are small today, and so, therefore, are one's chances of having a brother or sister with exactly the same immune characteristics. Fortunately it has been found that some HLA types are more important than others when matching and that even if you cannot get them all right, it is often sufficient to make a fit for only a few crucial types. In addition, drugs have been developed which can suppress the immune system and so limit organ rejection. As a result, transplants are now routinely performed and have saved countless lives. However, better immune suppressants are still needed, ones that will not damage a person's powers to fight infections, one of the gravest risks they currently face, particularly with bone marrow grafts.

But why is it necessary to go to such lengths in carrying out transplants? Why is there such extraordinary variation in individuals' HLA tissue types? Evolution surely did not do this to frustrate surgeons. So why does the human immune system contain such diversification? The answers to these questions came, once again, from the study of animals, in this case mice, which like other mammals share mankind's spectacularly varied immune markers. Scientists found that when they studied the mouse H2 system, which is an analogue of the human HLA system, they could detect intriguing differences in its powers to fight disease.

What they discovered was that a leukaemia virus triggered the condition only in mice that were of a certain H2 type. Some succumbed, some did not – and the crucial variable was their H2 markers. In addition, researchers found that mice of certain H2 types were capable of making antibodies only to a particular antigen. Some raised antibodies, some did not – and again the vital factor was the mouse's H2 markers. This observation clearly suggested why strains of mice could fight off leukaemia viruses and other disease agents – because they had immune systems that could tackle them before the ailment had taken hold of their body. A mouse's or a human's susceptibility depends on their particular H2 or HLA configuration, which explains why there are so many tissue types around. We have this staggering variety of white blood cell markers because each has been selected for its power to protect individuals from different types of infections. And of course, infectious diseases have been a major force in our evolution, and any slight improvement in ability to counter them – be it malaria, or smallpox, or tuberculosis or even the plague – will have accrued great success over generations.

In a sense our highly varied immune systems are the outcomes of deadly games of cat-and-mouse that have been played out, for millions of years, between the pathogens that pervade our environment and their victims. Once an attacking micro-organism is blocked by one variation in the immune system, it evolves to get round it. Then the immune system responds by evolving a new variation which can block that attack, and so on in a carousel of pathogenic thrust and immunological parry. The end product has been the creation of the human immune system, and its profusion of diverse HLA markers.

This system exists to recognize foreign invaders. That is how it protects us from infection, but of course that is also why it causes transplants to be rejected. The other side of this coin clearly must be that our immune system must learn to distinguish self from non-self, for if it did not we would reject all our own organs and tissues. Thus, as it develops, our immune system has to learn what is self, and what is foreign.

(There is one crucial exception to this law, by the way, and that concerns the failure of a mother to reject the growing foetus within her. It is, after all, made of foreign tissue, since it carries genes from its father. The foetus is not rejected because the cells of the placenta, which is of the same genotype as the baby, and which would be the first tissue to be attacked as a foreign graft because it surrounds the growing baby, do not display the usual HLA markers. The developing unborn child is in a sense hidden from its mother's immune system.)

Our body's defences are clearly exquisitely subtle and effective. However, as with any sophisticated system, their extreme complexity sometimes leads to a spectacular breakdown and often, when an individual's immune system fails, it turns on the body's own organs and tissue. This immunological 'revolt' is actually quite common, and its consequences can be extremely unpleasant.

An example is provided by the story of Liverpool-born Denise French. The first six years of her life were normal and healthy until, in 1960, her mother noticed that Denise seemed to be permanently thirsty, drinking tumbler after tumbler of water. Then Denise complained of feeling itchy and her mother found deposits on her vulva. The family doctor recognized these as crystals of sugar and through a simple test found that her urine was thick with glucose, a clear sign that her pancreas was not making insulin, the chemical which controls sugar levels in our bodies. In short, Denise had diabetes.

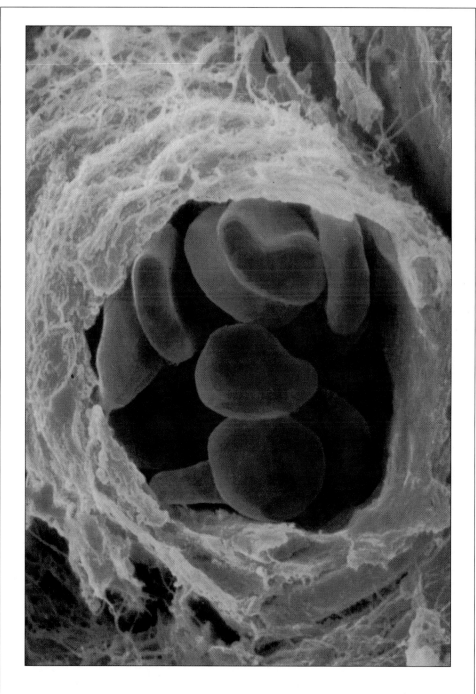

False-colour scanning electron micrograph of a group of red blood cells (erythrocytes). During its short life (about 4 months), a red blood cell covers about 15 kilometres every day for a total of 1,500 kilometres.

False-colour transmission electron micrograph of a section through mammalian skeletal muscle. The striated banding-pattern of the muscle can be seen. Skeletal muscle is responsible for voluntary movement of bones of the skeleton and of organs such as the eye.

Illustration based on a scanning electron micrograph of a human ovum, or egg, surrounded by numerous spermatozoa.

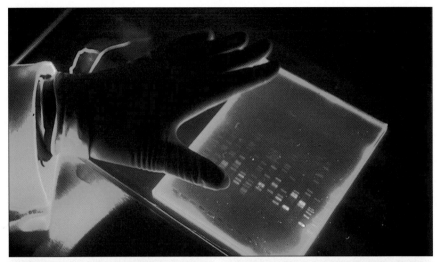

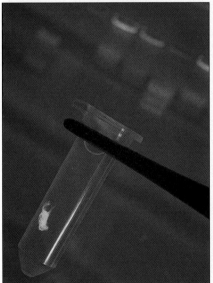

Top   *DNA sequencing by gel electrophoresis, a technique used to examine the base-pair sequence or chemical blueprint of lengths of DNA. The banding-pattern (fluorescent pink) is revealed under ultraviolet light.*

Left   *DNA fingerprinting. A sample tube containing a pellet of human DNA (white).*

Right   *Preparation of the agarose electrophoresis gel used to separate fragments of DNA into bands. Fragments of an individual's DNA are usually extracted from white blood cells, sperm or hair follicles.*

Denise was saved because it was possible to give her injections of insulin which, even for a six-year-old, became a natural way of life. Without them Denise would not have lived. But why had her pancreas stopped pumping out enough insulin? What had turned a cheerful, chubby child with shiny black hair into a tired, gaunt and bedraggled figure within a couple of months? The answer lay with her immune system which had attacked and destroyed those pancreatic cells which produce insulin. In other words, her own bodily defences were responsible for her condition.

But that explanation simply begs the next logical question: why had her life-saving immune defences turned into a potential killer? A clue was provided sixteen years after Denise's diagnosis when she noticed how thirsty her sister Sandra, eight years her junior, seemed to be. The sight of her young sister drinking tumbler after tumbler of water had an eerie familiarity to Denise's eyes. She gave Sandra a urine test kit, and sure enough, sugar was found to be building up in her body. Sandra, too, was suffering from the ravages of a misdirected immune system and soon was giving herself life-saving insulin injections for diabetes.

The story of Denise and Sandra is an important one, for the sisters became part of a study that showed the risk of getting juvenile diabetes (about one in 250) was critically linked to the HLA markers which defined their immune systems. Sandra and Denise shared the same HLA type, one which is many, many more times likely to be found in young diabetics than in the normal population. Through this and other similar studies, it was found that juvenile diabetes occurs in children such as Sandra and Denise, who are nearly all of type DR3 or DR4.

Such a disorder is called an auto-immune disease, a category of ailments that we now know includes many common illnesses such as rheumatoid arthritis, in which joints can become so swollen and inflamed that a victim can hardly even pick up a cup of tea. More than 80 per cent of victims have DR4 markers, which occur on average only among 25 per cent of the population.

Then there is the condition called ankylosing spondylitis, or more commonly poker spine, which occurs in about one in a thousand men, but is much rarer among women. The condition usually starts as a persistent lower back pain and degenerates until movements become extremely painful. It is an auto-immune disease in which the vertebrae of the spine fuse together, and victims are virtually all B27, an HLA type normally only present in about 4 or 5 per cent of the population. This strong association explains the familial inheritance of ankylosing spondylitis.

Another auto-immune disease is the condition of narcolepsy which affects about one in 100,000 and in which a person simply cannot control when he or she gets 'an attack of sleep'. They do not just drowse off, but are suddenly rendered unconscious. Virtually everyone who has this condition has the HLA type DR2.

These examples show that there is a wide, seemingly disparate range of diseases which are actually caused by attacks by the body's own immune system and that these disorders all have an inherited component involving genes on chromosome 6. Particular HLA types make people susceptible to auto-immune disease, although possessing a particular marker is insufficient on its own to cause a disorder. Only a small proportion of people with DR4 will get rheumatoid arthritis, for instance. Another factor, probably a viral infection, is involved in triggering attacks.

Clearly, white blood cells – with their involvement in transplants and rejections, as well as auto-immune disease – play a critical role within our bodies. This is putting it mildly. In fact, one type of white blood cell, the lymphocyte, is the heart and soul of the immune system. So let us look a little more closely at these cellular wonders before homing in on the solutions to the two main puzzles that have been posed in this chapter and which still have to be resolved: how does a body remember a previous viral or bacterial attack, and why do some people suffer from the predations of their own lymphocytes?

Lymphocytes get their name from 'lympha', which is Latin for 'clear spring water'. Lymph is the transparent fluid that circulates through special vessels in our bodies and it carries lymphocytes to all the nooks and crannies of the human frame. The lymphocytes' critical role in the immune system was recognized in the 1950s, when Peter Medawar and his colleagues showed that a mouse which had never received a skin graft, but which was given lymphocytes from one that had rejected a skin draft, would now itself reject a skin graft rapidly. In other words, the lymphocytes could transfer the immunity developed against the foreign skin graft from one animal to another. But serum without lymphocytes would not do this, an experiment that clearly implied that antibodies were not the mediators of skin graft rejection. It had to be the lymphocytes. The immune system must therefore have two components: one that is a producer of antibodies and the other that is cell-based and which is responsible for recognizing the foreignness of an infectious agent or poison. We now know that lymphocytes are involved in both categories. B lymphocytes make antibodies while T lymphocytes carry special markers that help them detect foreignness in virus-infected cells and eliminate them as well as carry out other tasks. In fact, there are two sorts of T-cells. One variety, known as the killer T-cell, eliminates virus-infected cells. The other sort are called helper T-cells and they send signals to B-cells telling them when to make antibodies. Understanding how these two systems of T and B lymphocytes combine has been an enormous challenge for molecular biologists and was achieved only through the intervention of two great scientists, Paul Ehrlich and Sir Macfarlane Burnet.

Ehrlich was born in 1854 into a Jewish family in a small town in Silesia, now part of Poland, and made his first major scientific contribution while still a medical student in Strasbourg. He realized that newly developed aniline dyes could be used to stain and identify blood cells on microscope slides, a discovery that led to the first full description of the different types of white blood cells and an understanding of their relationship to anaemias and leukaemias. Indeed, stains positively dominated Ehrlich's life. He was famous for leaving chemical blotches on his hands and clothes, on his laboratory tables, and even in the room of the inn where he lived. He was a gifted scientist nevertheless, and one of his most important ideas concerned antibody production. He argued that the cellular machinery for making an antibody must be in place before that cell encounters its corresponding antigen. In addition, he suggested, antibodies made by a cell are displayed on its surface, ready to catch their antigen victim as it hoves into view. When that antigen reacts with the antibody, the cell is stimulated to manufacture more antibodies. Surprisingly, however, Ehrlich himself never identified lymphocytes as the producers of antibodies.

It was left to Macfarlane Burnet, fifty years later, to fit the last piece into the jigsaw puzzle of our basic understanding of the human immune system's behaviour. He suggested, firstly, that lymphocytes are responsible for making antibodies, and secondly, that only one variety of B lymphocyte makes one variety of antibody. Basically, an antibody is a protein and when one of them, carried on the surface of its parent lymphocyte, recognizes its corresponding antigen, the former will latch on to the latter, a process that not only stimulates the cell to make more antibodies, but causes the lymphocyte to divide so there are even more cells to make even more of that particular antibody. When the body next encounters that antigen, there is therefore an expanded population of those lymphocytes ready to attack it, and only it. These lymphocytes should then produce all the antibodies necessary to counteract the infectious organism that is the carrier of the antigen. That is what provides the immune system with its 'memory', and of course that is the secret of that first vaccination which Jenner carried out two centuries ago. The cowpox antigens that had coursed through young James Phipps's arm eventually encountered lymphocytes which carried the specific antibodies to them. Those lymphocytes were then triggered into action and began to divide. As a result, when the smallpox antigens – which were indistinguishable from cowpox antigens as far as the body was concerned – arrived six weeks later, they were met by a well-prepared posse of stimulated lymphocytes and were promptly demolished before they could establish themselves in the body and wreak their dreadful havoc. Thus humanity has been protected, by similar medical subterfuges, against the ravages of count-less other ailments, a process that has been one of the prime creators of the current, healthy status of the Western world.

But there was still a major puzzle that Burnet had to solve. How could the body manufacture each of those millions of different lymphocytes, each tipped with its own highly specific spearhead? With so many different types of antibody to be produced, there are simply not enough genes in our DNA blueprints to create them all if each antibody had its own, unique, corre-sponding gene. Burnet's explanation was simple and brilliant. The genes responsible for the manufacture of the human antibodies must mutate dur-ing the development of the immune system. Millions of different mutations are produced, each in a different cell which, when it divides, produces its own particular form of mutated antibody. In short, our myriad different antibodies are produced by mutations that arise during the development of lymphocytes. In fact, we now know that there is not only a mutation in the way in which the antibody gene's individual DNA letters are reshuffled, but that the gene itself exists in many pieces, and that these fragments can be recombined in many different ways to create many different types of anti-body. Indeed, a B lymphocyte does not mature until it has recombined its antibody gene fragments in its own individual way. From this final distinc-tive version, all its daughter cells are then generated.

Now B and T lymphocytes carry out very different functions within the immune system. However, there are parallels in their behaviour. While a B lymphocyte makes antibodies, a T lymphocyte manufactures an antibody-like molecule called the T-cell receptor or TCR. However, unlike a B lymphocyte, which releases its antibodies into the blood once it has encoun-tered its antigen target, a T lymphocyte keeps its TCR locked on to its outer membrane. We can therefore think of T-cells as being armed with surface detectors which are used to search out antigens, such as those which appear on the surface of a virus-infected cell. Once these are found, the T-cell latches on to and kills the offending virus carrier.

These TCR surface detectors are rather myopic compared with their B lymphocyte counterparts, however. While an antibody can see fairly large chemical structures, including substantial portions of protein molecules, the TCR has a much more limited vision. It can only see small protein frag-ments, mostly eight to ten amino acids long, sometimes up to fifteen to twenty, but no more. And this is where the HLA molecule reveals its func-tion. X-ray crystallography studies have discovered that the HLA molecule acts like a clamp on a cell's surface, a biochemical vice that holds protein fragments for inspection by marauding T-cells and their surface detectors. These protein pieces are in turn made by the normal process of internal cell management, a form of biological husbandry in which proteins are regularly broken into fragments and cycled back to the cell surface where they are held in place by the HLA clamp. Normally, the protein pieces come from standard sources, from parts of the cell, and are not seen as foreign. These therefore provoke no attack from the T-cells. But if an unsuspecting virus has entered a cell, its protein components will also have been broken down into small pieces and, being clearly alien, will have been displayed like a microscopic beacon on which the T-cell can home in. And the T-cell which has the receptor with the right configuration to recognize a virus's alien protein fragment will then latch on to it and kill off the offending virus-infected cell, typically by punching it with holes through which poisonous enzymes are then injected.

A particular HLA type influences an immune response by acting as a better clamp for some protein fragments compared with other HLA types. People with such biochemical signatures will therefore be more likely to attract the attention of appropriate T-cells and so be better able to resist viral infections. That is a considerable advantage. However, if the HLA clamp is too good, it may fool a T-cell receptor into 'recognizing' a fragment that is from a normal protein from within one of its non-infected cells. The marauding T-cells will then mistakenly kill those cells. The result will be the severe tissue damage of an auto-immune disease – rheumatoid arthritis, ankylosing spondylitis, narcolepsy, or the debilitating loss of those insulin-making cells that struck Sandra and Denise.

Understanding how T-cells go about their predatory business and how their misdirected zeal causes such havoc to our tissue and organs has been one of the great triumphs of modern medicine. Today we can follow the paths they take when embarking on their bouts of tissue destruction, and can therefore begin to think about ways to prevent this happening. For instance, if we know the nature of the protein fragments that are displayed by HLA clamps and which stimulate the destruction of their cells, then it may be possible to find analogues of these fragments, ones that are not recognized by the T-cell. These could be used to clog up the HLA clamping machinery and prevent the destruction of their host cells, a technique that would halt the ravages of an auto-immune disease like diabetes without interfering with an immune system's normal operations.

This immunological awareness might even be broadened to include other forms of disease, for instance with the protein products that are made by mutated genes involved in cancers. Protein changes which occur during tumour formation may be recognized by T-cells and could form the basis, either for gingering up the immune system so it reacts with greater efficacy against a cancer, or even as an approach to vaccinating against it.

Another target is Aids, which is caused by the virus that almost completely eliminates helper T-cells. As a result, B cells no longer receive proper signals and no longer produce antibodies effectively. And without antibodies, victims succumb easily to infections caused by micro-organisms that would normally be mopped up by healthy immune systems. The answer is to eliminate the Aids virus before it eliminates the immune system, and the answer to that is vaccination.

There is still much more to be learnt about our exquisitely sensitive, highly regulated immune system, and its powerful, antigen-tracking systems. These have to be delicately balanced so they can distinguish between infectious organisms and normal uninfected tissues. The fact that we now understand the machinery of this complex process is largely due to the work of scientists such as Jenner, Medawar and Gibson, and Burnet. Greater advances lie ahead, through which we will make better and better vaccines for existing diseases, and develop new ones for as yet unconquered ailments such as Aids and, perhaps, cancer.

8

All in the Mind

Paediatrician Gillian Turner was working in an Australian clinic for the intellectually retarded in the 1960s when she stumbled on a discovery that was to have a profound impact, not just on genetics, but on the whole field of health care, and on the burgeoning field of neurology in particular. It remains one of the most remarkable stories of modern genetics.

Interested in the subject of X-linked mental handicap, the kind of cerebral impairment carried on the X chromosome and which affects only men, Dr Turner was preparing one evening to write a paper for a scientific meeting. She was staring at her files, which included photographs of male patients, when she complained to her husband that her cases were 'all so normal-looking'. But surely, she realised, looking normal in a mentally retarded population – in which individuals usually appear physically deformed in some way – was itself highly abnormal.

Intrigued, Dr Turner checked her clinic's photographs of male patients the next day and picked seventeen which she regarded as looking 'normal'. Then she looked at the clinic's records and found that eleven of these men had other family members, mainly brothers, who were also mentally retarded. It looked like a highly significant genetic association. 'It was a red letter day,' she recalls.

And so it was. Dr Turner had found the first indications that a form of retardation, then called Martin Bell syndrome, and since renamed fragile-X syndrome, and which was then thought to be extremely rare, is actually the most widespread form of inherited mental handicap in humans. It almost rivals Down's syndrome as the most common form of cerebral retardation, and surpasses all but cystic fibrosis as the most frequently occurring inherited disease among Westerners. (In fact, fragile-X syndrome is highly variable in the severity of its effects and some individuals are quite seriously retarded. They do not look 'normal' and tend to have long faces, square jaws and large testicles and can behave in wild, disturbed ways.)

However, it was not just the ubiquity of fragile-X that startled those scientists who began to follow up Dr Turner's work and to investigate the condition. It was its mode of transmission – sometimes from carrier females to affected males, but also, unlike any other known disease, from symptomless carrier males to affected females. And to cap this oddity came the discovery that from a quiescent, almost undetectable state, the syndrome would erupt, progressing rapidly in severity of intellectual impairment through two or three generations. Occasionally the reverse occurred, and the mental prowess of offspring would improve markedly from their parents' and grandparents'. The gene appeared to behave like a molecular pogo stick that bounced from generation to generation in unguessable directions. The subsequent unravelling of the underlying process responsible for this bizarre behaviour has proved to be one of the great scientific detective stories of modern times. It has revealed not only a previously unimagined genetic mechanism in operation but has also opened up one of the first molecular biological avenues into the brain.

Now, the mind is the most elusive of all physiological concepts, its secrets locked within the biological black boxes of our brains. Uncovering the contents of these neurological treasure chests is one of the last great challenges left to modern medicine and is pivotal to all attempts at alleviating the myriad illnesses that are produced when the mind malfunctions. Ailments such as schizophrenia, manic depression, alcoholism, senile dementia and other forms of mental disease still cause widespread anguish. Indeed, in the Western world, psychological conditions now absorb the largest slice of health service expenditure. The pressure to find cures is therefore unremitting.

The mapping and sequencing of all our genes – and up to half our total complement are thought to be involved in controlling brain function – will therefore be critical to this task. Indeed we should expect to learn not just about the sources of psychological impairment, but also about basic human individuality – the power to appreciate musical pitch, artistic ability, mathematical prowess and all the other variable manifestations of human brilliance. After all, it was mental ability that turned Homo sapiens from a struggling evolutionary bit player into ruler of our planet. Understanding how our brains helped us achieve that conquest should lead to a true philosophical appreciation of what it means to be human.

The hunt for the cause of fragile-X may seem a little mundane when considering such lofty goals, of course. Nevertheless the search is important, for it has already led to the discovery of one of the few genes that have so far been linked to intellectual activity. The fragile-X story therefore provides a general but illuminating insight into the difficulties and surprises that lie ahead in our exploration of 'the labyrinthine ways' of the human mind. More particularly, because the condition is usually so mild, it is more representative of the genetic effects that scientists are seeking in the field of mental abilities.

However, one of the first problems encountered by scientists was the simple problem of identifying individuals with the condition. The American geneticist Herbert Lubs had discovered that patients exhibited one striking feature – their cells had two little pieces of DNA that appeared to have broken, or were breaking, away from the long arms of their X chromosomes. These he called fragile sites and they have since given their name to the condition. Viewed through a microscope they appear as pairs of dots, like tiny colon marks, the detritus of some mysterious form of genetic grammar.

The trouble was that when other scientists tried to find these molecular punctuation marks, they could see no sign of them. 'It was something of a puzzle,' recalls Dr Grant Sutherland of the Adelaide Children's Hospital in South Australia. 'The sites just were not there when we looked for them.'

And there, in scientific limbo, Dr Lubs's discovery might have rested had it not been for the tenacious scientific sleuthing of Grant Sutherland. He began to scrutinize the methods employed at those few laboratories which had reported seeing fragile sites and then he compared them with techniques of his own and other laboratories. There was only one difference, he concluded after a protracted study. Fragile-successful laboratories were using an old form of the tissue culture medium, known as TC-199.

Tissue culture medium; it sounds imposing. In fact, it is simply a type of liquid food, a cellular ambrosia that is one of the standard tools of the modern molecular biologist. Made of sterile water to which salts, sugars, vitamins, animal blood serum and dozens of other ingredients are added, the medium provides a balanced diet for human cells (lymphocytes) and encourages them to grow and then divide. Without this food, no new generations of cells can be raised in the laboratory.

However, it is not replication itself that concerns cytogeneticists. They are really interested in the sequence of events that immediately precedes cell division. It is then that DNA, which has previously spread throughout a cell's nucleus, returns to form the tight little bundles that we see as chromosomes. Compressed to unique density, they become visible through microscopes for the only time in their lifecycles. Only at this stage is it possible to see, through microscopes, the flaws and gaps that are the stigmata of genetic and chromosomal disorders.

Effective supplies of tissue culture are therefore vital to researchers' work and this need has inculcated a desire on their part to make constant improvements to its efficacy. Unfortunately, with the culture media that had been developed as improvements to TC-199, scientists had clearly added something that actually reduced their power to see fragile sites.

'I used to get out the recipes for the different media and agonize and sweat over what might be the crucial discrepancy,' recalls Dr Sutherland. 'Then I discovered that if I added folic acid to the old tissue medium, the fragile sites disappeared from the microscope slides. And when I added folic acid inhibitors to the new culture media, the fragile sites appeared. Basically, everything stemmed from that discovery.'

Armed with new culture media free of folic acid, cytogenetics laboratories round the world, once blind to the faint signs of this ubiquitous ailment, could begin to search – and found evidence of the condition almost everywhere. Indeed, they began to discover it in places where it really had no right to be. Researchers started to find families who, when their pedigrees were traced, revealed grandfathers who looked and acted normally but who had clearly transmitted the condition to their daughters. These women were mildly affected, and produced children who were severely affected. Explaining that according to the rules of basic Mendelian genetics proved to be rather awkward.

Symptomless male carriers, sudden jumps in severity of the condition and the existence of affected females were certainly unexpected behaviours for an X-linked disease. Scientists were intrigued, and not just because of the condition's odd exotic behaviour, but because fragile-X was also so startlingly prevalent. Everywhere they looked, researchers found its telltale signs: patients in clinics with moderate mental retardation, and the X chromosomes of their cells bearing the disease's distinctive genetic colon marks. No race on Earth is untouched by the fragile-X syndrome, it now appears. About one in every 2,500 children born round the globe has the condition – which compares with the birth rates of Down's syndrome of one in 700, and cystic fibrosis of one in 2,000 in the West.

We know the causes of these latter two common conditions, added scientists, but what could possibly be the cause of the third? When first discovered, a simple, single gene lying somewhere along a person's X chromosome seemed the most probable culprit. The subsequent discovery of the syndrome's unpredictable behaviour challenged that assumption. Could it even be possible that several genes were involved? To many, it seemed the most feasible explanation.

But once again fragile-X proved it could still provide surprises, for in 1991 a global effort revealed that only one gene was involved, though it had some strange assistance. In fact, the 'breakthrough' – and although they dislike the word, scientists do use it in this case – came in two parts. The first was the work of several groups, including teams led by Dr Ben Oostra in Holland, and Dr Tom Caskey of the University of Texas in Houston.

These groups, by laborious 'reverse genetics', traced the source of fragile-X to a gene on the X chromosome called FMR-1. When the message of that gene – and as yet no one knows what is its protein product – is disrupted then the symptoms of fragile-X are produced, they concluded.

But what could possibly disrupt it, doctors wondered? The answer was not obvious and once again science might have been stumped had it not been for the industrious Dr Sutherland. His own parallel researches revealed there was a triplet of chemical bases, two cytosines and a guanine, that lay right beside the FMR-1 gene on the X chromosome, and more to the point, this CCG repeated itself in a way that paralleled the symptoms of the disease. In simple terms, the more copies of CCGs beside the FMR-1, the more severe are symptoms of fragile-X.

In a normal individual, this triplet is repeated between six to sixty times without causing problems, producing just another of those strange, repetitious, seemingly functionless parts of the humane genome. In symptomless, carrier males – or 'transmitting' males as they are more usually called – there are 70 to 200 CCG sequences. And in those suffering from the full-blown syndrome it is repeated hundreds, sometimes thousands, of times. This process is called 'genetic amplification' (not to be confused with DNA amplification, or polymerase chain reaction, which we discussed in Chapter Four).

Genetic amplification therefore offers a neat explanation for the fragile-X's peculiar passage through the generations. The CCG repeats in some families are larger than average – probably around 100 to 200. At this magnitude, the region becomes unstable and occasionally, as it passes from generation to generation, it rapidly jumps in size and in doing so produces the symptoms of fragile-X syndrome. The repeat interferes with the FMR-1 making its protein message. The gene can tolerate a certain increase in the repeat's size and still make a message that works. Once it gets beyond that, however, the gene cannot cope any more.

Exactly why long repeats cause fragile-X or what protein function is interrupted has yet to be found out. Nor is it clear what are the pathways by which this genetic malfunction takes its effect on the intellect of victims. Nor is it understood how the condition is passed by transmitting males to affected females. 'This is not an intelligence gene,' says Dr Caskey. 'It seems to affect the development of the foetus in some way we have not worked out. It is clearly very complex in action, and also in malfunction. It is going to be intriguing work to unravel what is going on.'

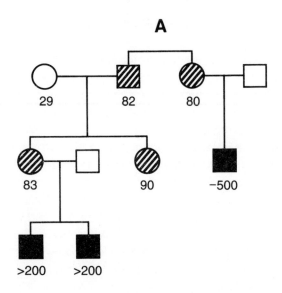

The inheritance of fragile-X. The shaded individuals at the top of the diagram represent a transmitting brother and sister. Both have an unstable number of CCG repeats (82 and 80 respectively). The sister (shaded circle) subsequently gives birth to a son who has several hundred repeats – and fragile-X syndrome. The brother (shaded square) has two daughters both with unstable regions, and one of these women goes on to produce two sons, each with hundreds of repeats – and fragile-X.

(In fact, Sutherland, Caskey and the other fragile-X researchers had stumbled on a phenomenon that was to prove to be surprisingly common – as we saw in Chapter Five in which there is a description of how triplet repeats were found within the gene that causes Huntington's chorea. In this case, it was a repeat of cytosine, adenine and guanine – CAG – which caused the disease. Normal individuals have 11 to 34 CAG repeats. Victims have between 40 and 100, an excess that must distort the business of coding, so that a harmless protein – whose identity still remains a mystery – becomes one with a cumulative toxic effect. In other words, the triplet repeat lies within the gene. With fragile-X, however, the triplets lie just outside the FMR-1 gene and interfere with its work only when the number of CCGs becomes excessive, preventing a normal protein product being made in some way. In addition, by 1993, two other inherited diseases had been linked to triplet repeats – myotonic dystrophy in which excess of CTGs (a cytosine, a thymine and a guanine) cause the problems, and spinobulbar muscular atrophy, which has been linked to CAG repeats.)

But scientists are certainly not daunted by the task ahead. Twenty years ago no one had even heard of fragile-X. Now its ubiquitous tendencies have been uncovered and already a tentative mechanism has been established for its strange passage from generation to generation, which is a fairly creditable achievement.

And there are other clues about the nature of the genes that control our intellect. Many of these have come, not from studying conditions where a gene has gone wrong, but from disorders where there are too many, or too few, normal copies. A classic example is provided by Down's syndrome. Children with this malady suffer, among many symptoms: flat faces, poor muscle tone, short hands with stubby fingers and marked susceptibility to infections. Most pronounced of all, though, is the severe mental retardation that frequently afflicts a Down's syndrome child. Now all these features have one common cause. Instead of having 46 chromosomes, most Down's syndrome children have 47. Due to an error during the meiotic division that creates either its sperm or its egg cell, an extra copy of chromosome 21 is inserted. And that is all. A victim has three, not two copies, of chromosome 21. It does not sound much, yet this slight excess in chromosomal cargo is enough to produce the gross distortions of intellect, as well as the severe physiological problems, that are typical of the condition.

In other words, a difference which constitutes less than 2 per cent of the chromosomes put together can have a gross effect on intellectual capacity. We can therefore see that having the correct balance of genetic material, not too much and not too little, is as important as having DNA that is error free. This dosage effect provides scientists with a critical guide for associating certain genes on chromosome 21 with the development of intellect. It is an important lead.

Nevertheless, those who are seeking signs that the genetic make-up of the mind is simply going to fall apart under the scrutiny of molecular biologists will not find total comfort from these examples. There are too many twists and turns to suggest that the job is ever going to be straightforward. So at this stage perhaps we should consider just what kind of evidence there is for believing our thoughts are channelled at the behest of our genes.

At first glance that may seem an absurdly unnecessary step. We inherit 50,000 to 100,000 genes from our mothers and fathers and about half of these are thought to be responsible purely for brain development – which indicates that our capacity to deduce, make rational decisions, memorize crucial lessons and take control of our immediate surroundings probably has a huge genetic component. But are these powers completely inherited, or is there room for environmental influence? Is it nature or nurture that is the final mediating influence on our minds? The answer is both, of course.

However, it is the *raison d'être* of this book to look at human physiology in molecular biological terms. So from this perspective let us examine the operation of the brain. Or at least let us try to. The trouble is there are few absolutely clear-cut ways of doing so, the strands of inherited traits being so tangled with those that come from our environments. After all, how can we really tell if we get our temperaments from our parents' genes or if we derive them from apeing their behaviours from birth until these actions become so fixed in our psyches that they appear as fundamental to our natures as the colours of our hair and eyes?

The obvious way would be to look at two genetically identical individuals with quite different backgrounds. After all, if two people have exactly the same complement of genes, it follows that observed differences between them must be due to differing environments – family, schooling, culture, geography, and countless other variables. And, as we have seen, there is only one type of person that falls into this category – identical twins.

Unfortunately for science, identical twins are rarely raised in dissimilar ways. Indeed, they are usually brought up in a manner that stresses their indistinguishable physiques. They are dressed in the same clothes, given similar forenames and treated almost as one person. Unravelling the strands of nature from the straws of nurture of such pairs of doppelgängers has therefore proved extremely difficult.

In a few, rare cases, however, identical twins are separated at birth and adopted by different families. They offer a clear-cut opportunity to find out if we are shaped by the genes of our parents, grandparents and ancestors, or by the manner in which we are moulded, after birth, by external events. Unfortunately there are probably only a few thousand reared-apart identical twins in the Western world, so they do not represent a colossal database. Nevertheless they are an extremely informative one, which explains why scientists prize them so highly.

One group that has committed years to the study of twins is led by Thomas Bouchard, professor of psychology at the University of Minnesota and founder of its twin studies project. To date, Professor Bouchard and his team have tracked down and studied more than a hundred separated identical twins – the largest single collection in the world. And from their disrupted lives Bouchard has unravelled a surprising picture of the inclinations which we seem to inherit and pass on to our children.

At Minnesota, once a pair of separated twins have been traced and agree to co-operate (and nearly all do), they are flown, expenses-paid, to the university. For six days, personalities and intelligences are assessed, exhaustive physical examinations are gone through, and a total of 15,000 questions asked about their interests, values and beliefs, as well as their dietary and recreational habits, their sex lives, allergies, dental records – even their TV-viewing preferences. Then results are compared. For each test and questionnaire, the values of one twin are contrasted with his or her partner to produce a figure known as a correlation. The greater the similarity in their test scores, the higher the correlation.

Take physical features such as height, fingerprinting and hair colour. They show an overwhelming similarity. Correlations are typically around the 0.9 (90 per cent) level. After twenty, thirty, forty, even fifty years of leading different lives, nearly all reunited twins still look alike. That may not be surprising in itself. However, cases often tend to be accompanied by more unexpected similarities when more disparate variables such as personality, brainwave pattern, IQ, even weight, are assessed. Then the data becomes even more intriguing – for these features are also found to have quite sharp correlations, between 0.5 and 0.7 (50 to 70 per cent). A good anecdotal example is provided by Jim Springer and Jim Lewis. When brought together, after thirty-nine years of leading utterly separate lives, they looked superficially dissimilar, sporting different hairstyles and clothes. But after their first few hours together, their reminiscences began to peel back layers of lives that contained some astonishing parallels. For a start, they both drove Chevrolets, chain-smoked, gnawed their fingernails and enjoyed stock-car racing. Each had built a workshop for himself in the basement of his house. In his, Jim Lewis built miniature picnic tables, and in his, Jim Springer constructed miniature rocking chairs. Furthermore, each had been married twice, and both of their first wives were named Linda, while both of their second wives were called Betty.

To many, these coincidences are unsettling, partly because they seem to refute a basic tenet of human existence – that there is a sanctity to the individual; a belief that we are all special in a discrete way. Genetically interchangeable, and manifestly alike, twins like Jim Lewis and Jim Springer disturb that comfortable notion. The fact that their lives unconsciously followed paths of such bizarre similarity only heightens that unease.

Of course, some features shared by the two Jims could be nothing but accidents. It is inconceivable that anyone could be genetically programmed to fall in love with persons of a certain name (no matter what Oscar Wilde argued). Similarly, there is no way they could have been biologically 'hardwired' from birth to end up making workshops so they could manufacture wooden furniture. Nevertheless it is also obvious that although the two had spent all their lives apart, they acted in many ways like mirror images of each other, an observation supported by personality analysis. When the two Jims were given tests that measured tolerance, conformity, flexibility, self-control and sociability, their results were so close that they could have come from the same person. Yet, of course, both men had led utterly different lives.

Even more surprising are separated twins' similarities in religious beliefs, jobs and political persuasion; features of our lives we would instinctively assume to be environmentally mediated. Surely we choose our jobs with due deliberation; think meticulously about our religious beliefs; and only plump for a political party after carefully considering its policies? Not always, it seems, for twin studies suggest our genes have sent us halfway down the road towards our choice even before we had begun such deliberations. As another twin researcher, Richard Rose of Indiana University, has put it: 'No dimension of behaviour seems to be immune to the effects of genetic expression.'

The story of Mark Newman, a member of the Paramus Volunteer Fire Department in New Jersey, illustrates some of these points. His colleagues were baffled when, attending a convention, they saw another firefighter, Gerald Levy, who looked startlingly like Mark, right down to his carefully trimmed moustache and wire-rimmed glasses. They persuaded the double images to meet. 'I talked to the reflection and it talked back to me,' recalls Mark. 'We're clone-heads,' was Jerry's more cryptic reaction. Born in Manhattan in 1954, and separated when five days old, neither had known he had a twin brother. Despite their different backgrounds, however, both were bachelors, keen hunters and enthusiastic motorcyclists. But most importantly, they shared an intense interest in firefighting, their chosen livelihood.

Of course, the notion that they were genetically programmed to become firemen is simply counter-intuitive. Most people's experiences are of picking jobs in a thoroughly haphazard way, from the small boy who suddenly wants to be a train driver, to the middle-aged businessman who abruptly abandons his career to be a humble cleric. Yet when twins are asked to select their preferences from a long list of various occupations, and to rate the importance of statements about job satisfaction such as 'being busy all the time' and 'the chance to do things for others', results consistently show that each pair of twins – regardless of upbringing – has a correlation of about 0.3 (30 per cent). What is being measured here is not preoccupation with a particular job, more a reflection of the intrinsic importance that a fulfilling job has to an individual. And if Mark and Jerry had very similar views, rating job satisfaction high above wages or personal safety, then their choice of firefighting seems less surprising in those circumstances. Their common inheritance merely narrowed down their list of occupational options.

We should proceed with care with such ideas, however, for the data most certainly do not rule out the role of the environment or of education and only suggest very broad, indistinct predispositions. Twin research indicates we are influenced by our genes, but are certainly not prisoners of them. In fact, it is more a matter of identical hard-wiring. The eerie coincidences uncovered by these studies stem from the duplicate sensory systems of genetically identical individuals. Each selects a similar world in response to their inherited cues. Some people merely inherit one or two similar attributes – such as an inability to taste PTC, or a failure to distinguish the colours red and green – which links their outlook on the world. Twins are born with a whole of array of identical perceptions and this must greatly narrow down the list of their likely behaviours, producing some of the remarkable coincidences outlined above. As Professor Bouchard puts it: 'There are lots of shaving creams that make my skin itch. It shouldn't be that surprising if people with the same skin pick the same cream that suits them.'

To some people such an analysis no doubt sounds like biological determinism, a doctrine which implies we are helpless prisoners of our genes and that we lack the free will to change and improve ourselves. This is not true. For one thing, twin studies tell us nothing about the action of individual genes, they only indicate broad, indistinct categories of inherited behaviour. As Arno Motulsky puts it: 'The biological substrate is still a black box.'

And of course, there are criticisms to be made of twin studies. For one thing, people who adopt children tend to come from the same socio-economic and educational backgrounds. Separated twins may therefore be raised in unconsciously similar ways which would skew results. However, respond researchers, if such criticism were true, close correlations would also appear in tests between twins and their adopting parents. This is not the case.

When twin research began in earnest in the 1960s and 70s, it was expected it would uncover some traits that were almost purely inherited – such as impulsiveness, a tendency that seems to affect people from birth. On the other hand, it was also anticipated that other inclinations, such as motivation and the drive to achieve, would be found to be fuelled totally through family influence. This has not been found. Nearly all facets of behaviour have a significant, but not necessarily overriding, genetic baseline, it seems.

However, most persuasive of all are the results of personality tests carried out on identical twins reared apart, and those brought up together. At Minnesota, the average correlation figure produced by reared-apart twins was 0.49, and for those brought up together it was 0.52 – a negligible difference. Family background, similar or dissimilar, made virtually no impact, suggesting a fairly crucial role for our genes in establishing the type of person we are – and a very small one for domestic environment.

'The fact that inherited inclinations are as extensive and as general as we have found seems to be saying something special about the way human beings are constructed,' says Professor Bouchard. This special message, he argues, is that people inherit, and pass on, highly variable intellectual powers and motivational attributes quite specifically to maintain a deep pool of very different human beings. This process was important in our evolutionary past but has now taken on a critical significance as it helps drive the accelerating pace of modern cultural change. Without this variation we would all be more uniformly average, says Bouchard, and would live in very different types of society.

And this last point raises a crucial issue. Twin studies clearly show that genes produce powerful variations in mental attributes. But they do so in individuals, not between races, or between classes, or other general groupings of human beings. As we shall see in the next chapter, one of the most surprising and satisfying discoveries that has been made in modern genetics is the fact that most inherited variation operates at an individual level. Genetic variation between races is extremely small in comparison. It is estimated that about 85 per cent of human variation comes from differences between different individuals from the same country, be they Britons, Nigerians or Taiwanese. By comparison, variation between races – say Europeans and Africans – is about 5 per cent. We can make virtually no accurate predictions about a person's abilities purely from his or her race. Individuals, not nations, are the main repository of human variation. The discoveries by twin researchers support this view, and should therefore not be construed as endorsing the redundant notion that there are meaningful racial differences in intellectual ability.

Now the mental powers that we have discussed encompass many different categories of the mind. As we have seen, there is the issue of personality which defines the type of human being we are – outgoing, emotional, shy, aggressive, or whatever. There is intellect, which sums up our power to reason logically, and then there is the quite separate issue of individual powers such as musical ability, manual dexterity, spatial prowess and others. What then can genetics and molecular biology say about the details of these different facets of the mind? Well, it can say something important now, but will indisputably say a great deal more in the near future.

Let us take the question of personality first. What is quite clear from the work of Thomas Bouchard, Richard Rose and others is that the idea that only a few genes stamp out human character is an incorrect one. There must be a minimum of at least a hundred, perhaps several hundred genes, of as yet unguessable function, which cumulatively shape personality – in conjunction with environmental influences. Individually, none of these genes plays an overwhelming role within the mind. Instead, each contributes a small amount to the mental mosaic of personality.

In short, there is no gene for personality. Instead, there is a large, complex group whose individual members add only a slight brush of colour to the painting of our characters. Occasionally, though, a single gene, or small set of them, upsets this delicate image. The effect is like throwing an entire pot of emulsion over the delicate canvas of the mind, smothering the fragile picture that has been painted on it, and producing the gross effects of mental illness – like the wild mood swings of manic depression, or the deluded, irrational thoughts of schizophrenia.

Indeed, with their simple aetiology and clear symptoms, these two conditions which so savagely disrupt the calm sanctity of personality and which cause such misery within families have seemed particularly ripe for the attentions of the modern molecular biologists, armed with their rflp probes and other accoutrements for uncovering genetic linkage. Let us look at schizophrenia first. The ailment, according to a *Nature* editorial in 1988, is 'arguably the worst disease affecting mankind'. Now that seems an extreme claim. Doesn't malaria kill a million children a year; doesn't Aids affect hundreds of thousands of individuals? Schizophrenia is not even fatal. Yet the assertion has some validity, for the disease does afflict 1 in 100 people in the West and, as the editorial points out, victims 'must limp through a normal lifespan incapable of functioning as normal people.' And its symptoms are certainly terrible – visions, voices, delusions of persecution and many other indications of a serious disruption of personality. (The term schizophrenia means disintegration of the mind, by the way, and has nothing to do with split personality, as is popularly assumed.)

Take the following example, outlined by David Stafford-Clark and Paul Bridges in their textbook *Psychiatry for Students*, of a boy who had a trouble-free childhood and went on to success at school where he was clever and popular. 'When aged fourteen years he became aggressive and difficult. He was referred to a child psychiatrist. After a few months he was admitted to an adolescent unit but his condition deteriorated and so he stayed in that unit until he was sixteen years old. At that time, he was transferred to a long-stay ward.' Later the boy absconded and returned to his parents, but he was so unreasonable and hostile at home that he had to be returned to hospital where he has remained. 'He is now twenty-eight years old,' state Stafford-Clark and Bridges. 'During the early years of his admission he was often argumentative, aggressive and difficult in behaviour, but he could communicate fairly readily even though he tended periodically to maintain a delusion that his medication was deliberately intended to harm him.' The two psychiatrists conclude: 'Unless there is a major remission, whether through current or future treatment or spontaneously, he will never leave hospital.'

It is an all too familiar story for the psychologist; young lives ruined and parents left to try to cope with grief and feelings of guilt. As Daniel Weinburger of the US National Institute of Mental Health puts it:

'Schizophrenia seems to involve impairment of what we think of as the highest psychological functions, the most intricate, sophisticated, complex psychological functions people have – those aspects that separate the human species from the rest of the animal kingdom.' But what is causing this disruptive impairment? What is the nature of the neurological pot of paint that is being thrown over the delicate tracery of the human mind?

These are questions over which psychologists and psychiatrists have sweated, debated and argued for more than a century, though twin and family studies clearly show some strong inherited component. It is estimated that in contrast to a 1 per cent risk of schizophrenia in the general population, the danger for members of a first-degree relative with the disease is 10 per cent and for an identical twin so diagnosed it is 50 per cent. Like the mark of Cain, the disease seems to blight certain families. Indeed, some pedigrees even look like those of families affected by simple, dominantly inherited illnesses – but only superficially. Closer examination usually reveals a more complex picture. 'Skipping of generations and other irregularities are the rule rather than the exception in families multiply affected by schizophrenia and indeed about two-thirds of schizophrenic patients appear to have no family history of the disorder at all,' says Peter McGuffin, professor of psychiatry at the University of Wales Medical College in Cardiff.

Yet in 1988 the world was briefly led to believe that such complications were a mere illusion and that a single gene which predisposes people to schizophrenia had been discovered. The claim was made by Dr Hugh Gurling and Dr Robin Sherrington of University College and Middlesex School of Medicine in London. They had examined seven British and Icelandic families with schizophrenia among their members, and professed to have seen 'strong evidence' of a genetic defect on chromosome 5 which inclined relatives to the disease.

The trouble was that the paper outlining the discovery, which was announced in *Nature* on 10 November 1988, and which triggered the sensational editorial described above (as well as worldwide press reports), was accompanied by a paper from a second group, led by Kenneth Kidd of Yale University School of Medicine, who reported finding no evidence at all to support the claims of Gurling and Sherrington. For researchers who had announced that they had found 'the first concrete evidence for a genetic basis to schizophrenia', this early refutation was not an auspicious start. And then, when more teams tried to replicate their findings – using different markers situated along the same stretch of chromosome 5 – they too reported that they could find no link with the region and the occurrence of schizophrenia.

'Gurling and Sherrington did not cook the books,' says Professor Robin

Murray of the Institute of Psychiatry in London. 'Their positive findings arose by chance, but they and other molecular geneticists did not realize that the simple statistical techniques, which had always been appropriate for medical disorders, were not able to cope with the complexities of schizophrenia.'

In fact the study by Gurling and Sherrington turned out to be part of a 'false dawn' for new genetics and psychiatry, as Professor Murray puts it. An equally headline-grabbing example was provided by Janice Egeland's 1987 study of manic depression. Indeed, in many ways her work was even more dramatic than Gurling and Sherrington's, for her research was carried out among the strange and romantic people of the Old Order Amish, who live in Lancaster County in Pennsylvania.

Now the Amish are a 'geneticist's dream', as Victor McKusick describes them. An archly conservative branch of the Mennonite Church that fragmented from the main church in the 1600s and which subsequently emigrated to the United States, the Amish today are an anachronism isolated in the midst of the high-tech world of modern America. Most still use nineteenth-century implements for farming (the main occupation) and reject cars, telephones and other twentieth-century machines as disruptive influences on community life. They live quiet, religious, highly co-operative lives, usually marry locally and can invariably trace their descent from the original founders of the order, thanks to the excellent family records that they keep. Most Amish are related to each other several times over. Indeed, more than three-quarters of their number are accounted for by only six surnames. No wonder McKusick views them as a genetic gold mine.

However, it took Egeland, one of McKusick's students, more than fifteen years to appreciate fully the accuracy of that description. In 1959 she had immersed herself among the Amish in order to carry out sociological research, but in the mid-1970s, as the first linkage and genetic studies were being launched in other medical fields, she realized she might be able to carry out similar work with the Amish. Her talks with community leaders had uncovered a dichotomy of 'strong' and 'weak' families, as they were known locally, the latter group being afflicted by a much higher preponderance of emotional problems, ranging from simply arguing with preachers to committing suicide. Such cases have the hallmarks of manic depression, a severe, debilitating condition that produces wild swings of mood from elation or hyperactivity to utter misery and despair.

We all experience fluctuations in our feelings, of course. A tiff with a spouse or a pleasant meal with friends usually triggers feelings of guilt or a sense of well-being. The symptoms of manic depression go well beyond such reactive sensations, however. They are unrelated to day-to-day encounters and erupt like a tidal wave of emotions released by some unknown, internal biochemical explosion. At its crest, thoughts, ideas and wild impulses rush through the mind in an overwhelming barrage. At its trough, loneliness, gloom and black depression flood the senses, producing, at its worst, a despair of such intensity that about 10 per cent of victims end by killing themselves.

And as with schizophrenia, psychiatrists have long suspected there is a tendency for the condition to run in families. There is an 80 per cent risk of succumbing to the ailment if a person has an identical twin who is manic depressive, for instance, while that risk is reduced to only 23 per cent if the twin is fraternal. Less rigorous, but more vivid support is also provided by historical figures. 'I inherited a vile melancholy from my father, which has made me mad all my life,' wrote depressive Samuel Johnson. Then there was Vincent Van Gogh whose mother's family was riddled with depression and who appears to have passed it not just to the painter, but to several of his siblings. Vincent's sister Wilhelmina spent half her life in an asylum, while his youngest brother Cornelius committed suicide aged twenty-three. It is 'our neurosis, a fatal inheritance', acknowledged Van Gogh, who died in 1890, aged thirty-seven, from the complications of a self-inflicted gunshot wound.

The third and most intriguing example is provided by Lord Castlereagh, the depressive Tory politician who cut his own throat in 1822. The prospect of inheriting Castlereagh's melancholic madness completely unnerved his relatives, especially his nephew Captain Robert FitzRoy. As a result, when FitzRoy was given captaincy of the HMS *Beagle*, he insisted on paying to have a gentleman companion, and selected Charles Darwin, to keep him from despondency on the five-year journey. Darwin later recalled FitzRoy's frequent 'low spirits, on one occasion bordering on insanity' during the voyage. It is fascinating to realize that Darwin's odyssey – which led directly to the overturning of man's self-glorying image of himself, and indirectly to the creation of modern genetics – only came about because of another man's dread of inherited insanity. As Stephen Jay Gould points out, had Castlereagh not committed suicide, his nephew would probably not have worried about family madness and would not have sought a learned companion to distract him. The history of science would then have been very different. (Ironically, FitzRoy died aged sixty in 1865 – when he cut his throat. 'His end was a melancholy one, namely suicide,' Darwin wrote, 'exactly like that of his uncle Lord Castlereagh, whom he resembled closely in manner and appearance.')

Such tales, however graphic in content, prove nothing, of course. More rigorous testimony is required to establish a link between genes and depression. And that is why the Amish looked so promising to Egeland and others.

Robert FitzRoy.

Apart from genealogical punctiliousness, the Amish do not drink or take drugs and this abstinence makes it easier to spot and diagnose manic depression among them. In other societies, drug abuse and alcoholism often mask the symptoms of the disease. So Egeland began to look at those families among whom suicides ran 'in the blood'. Perhaps among these, she thought, she could find clues to the biochemical trigger which floods the mind in such an emotionally devastating manner.

Egeland began by selecting four large, extended families of whose members twenty-six had variously hung, shot or drowned themselves over the previous hundred years. (It should be stressed that Amish suicide rates are actually relatively low, although numbers worried community leaders enough for them to want to co-operate with the project.) And by the mid-1980s she had amassed voluminous files of data about these families and members who were diagnosed as suffering from manic depression. She used that information to draw a tree that appeared to demonstrate pretty conclusively that a gene, or cluster of genes, was producing the condition. But how exactly could she track it down? She asked the help of researchers at Massachusetts Institute of Technology, and under their guidance she began her 'blooding', taking samples from a total of eighty-one related Amish people including nineteen manic depressives.

The researchers then ran twenty different sets of probes over these samples to see if any appeared only among the genes of manic depressives. All but one produced negative results. So the team discarded these and concentrated on their single, promising probe – on the short arm of chromosome 11. This data seemed strong and the team calculated the odds that the link between their region and manic depression was merely due to chance was 1,000 to one. Egeland and her colleagues wrote up their results, reporting that 'a dominant gene conferring a strong predisposition to manic depressive illness' had been found. It was published in *Nature*. However, as was to be repeated later for schizophrenia, *Nature* also reported two other teams who had found no link with manic depression when they ran the probes over the region .

Not a problem, said the researchers. These results were simply a matter of heterogeneity. In other words, there must be several genes which cause manic depressive illness. The Amish just happened to have a type that was not found in those families tested in the other studies.

But Janice Egeland was a scrupulous researcher, and she continued to study her families and extend her Amish pedigree. In 1989, she noted that two people who had been included in her original survey, and who had been diagnosed as being healthy, later succumbed to mental illness. The discovery upset the delicate connection that she had made between the disease and chromosome 11 and the tight linkage she had established began to unravel into a limp mess of weak correlations.

But there was worse to come. John Edwards, professor of genetics at Oxford, provided the *coup de grâce* in a devastating essay in *Psychological Medicine*. If you saw an archer fire a single arrow and hit the bull, you would be impressed, he pointed out. But if he fired twenty arrows, and only hit once, you would not be. And that is analogous to Egeland's study, he said. The team had actually run twenty markers through the Amish pedigree, and ignored nineteen because they produced negative correlations. In fact, those discarded markers should have been included in the statistical analysis of the Amish group – and that would have reduced its significance twentyfold, argued Edwards.

Today, both the Amish, and the chromosome 5 study of schizophrenia, stand as honest but discredited pieces of research, victims of the false dawn through which have passed both the new genetics and the study of mental illness. Multiple causes of single psychological disorders; lack of large pedigrees; misdiagnoses; wrong statistical methodology and the baffling complexity of emotional illness in general have all combined to defeat researchers. That does not mean the cause is fruitless, of course, it merely shows that the battle will be a lot tougher than it was first expected to be.

It was a point clearly illustrated with another controversial study that combined human behaviour with our genes. In this case, the research was carried out by a team led by Dean Hamer, at the National Cancer Institute, and it seemed to associate homosexuality with a piece of DNA found on the long arm of the X chromosome. In the families studied by Hamer, it was discovered that if one version was inherited, then men in those families tended to be homosexual. It was a well-conducted study, scientists admitted – and that was enough. When Hamer's paper appeared in *Science* in July 1993, he was hailed as the discoverer of 'the gay gene', generating banner headline stories round the world.

In fact, Hamer had only narrowed down the location of this piece of DNA to a region of about 2 million base pairs. Nor did Hamer claim that his mystery 'gene' was the gay gene. Indeed, he was at great pains to stress the opposite. 'We have not found the gene, which we don't think exists, for sexual orientation,' he said at the time. 'We know there are some gay men that don't carry this region, and there may be heterosexual men that do. We have found something that influences sexual orientation without necessarily determining it.'

One crucial point was that in his study Hamer had used only 'exclusive homosexuals', i.e. gays who were interested only in other men, and who were not in any way bisexual. In fact, the majority of men who have had gay experiences (and Hamer's study concerned only male homosexuals) have also had sexual relations with women and so fall outside the study's extreme definition. Hamer's research did not encompass a large number of gays who find both men and women sexually attractive.

And that is the nub of the issue. Sexual preferences come in a broad array, from the exclusive homosexual, to the heterosexual who has gay forays, to the heterosexual who only finds members of the opposite sex attractive. The idea that a single gene could control these widely varying reactions is ridiculous. Nor, as we have seen, did Hamer condone the notion in any way. His caution tended to be lost in the furore, however. If nothing else, the story illustrates the difficulties inherent in linking human behaviour to our genes.

In fact, the problems we are now facing in sorting out the genetics of the mind are similar to the difficulties that faced Gregor Mendel's predecessors who vainly attempted to uncover the basic laws of inheritance. They looked at characteristics that controlled economic importance and overall yield, for example, and did crosses between widely disparate varieties. In short, they picked a problem that was too difficult to solve in one go. It took the genius of Gregor Mendel to decide to pick simple, well-defined characteristics that would yield clear patterns of inheritance.

So our approach to the study of the mind must be to dissect out specific components of behaviour and ability that are well-defined like the smooth and wrinkled peas Mendel worked with. Overall personality, or whatever it is that is measured by an IQ test, comprises enormously complicated attributes that will be influenced by many different genes, and by the environment, in ways that cannot be sorted out one at a time.

Of course, taking this simplified approach is easier said than done, as we have seen from the examples of schizophrenia and manic depression, two well-studied psychiatric illnesses which have still to reveal their genetic secrets. Nevertheless there are hopeful avenues, including schizophrenia and manic depression which, for all the blind alleys they have led us down, are still being studied because they represent important routes for exploring how our DNA influences our thinking processes.

And it would be wrong to think that modern psychology has not reaped any benefits of the new genetics so far. In the arena of intellect, we have already learned about one breakthrough – the discovery of the fragile-X gene. And there is another, one that concerns an ailment which threatens to engulf the ageing populations of the West – Alzheimer's disease.

A degenerative illness of the central nervous system, Alzheimer's is an inexorable destroyer of cognition and is the principal cause of dementia in the elderly. As more people live longer, more individuals are entering the danger zone for Alzheimer's, creating a vast, silent epidemic that yearly destroys millions of lives. Having survived a succession of life's challenges, victims have to endure the loss of the most human of attributes: reasoning, abstraction, language and memory. 'Such a fate now awaits millions of individuals in races and ethnic groups worldwide,' says Dennis Selkoe, professor of neurology and neuroscience at Harvard Medical School. 'The dramatic rise of life expectancy during this century, primarily through the cure of infectious diseases, has enabled many of us to reach an age at which degenerative diseases of the brain – particularly Alzheimer's disease – become common.'

In hard statistics, 5 per cent of those aged over sixty-five years, and 20 per cent of those over eighty have Alzheimer's. Translated into numbers, there are 300,000 Alzheimer's victims in Britain, and in the United States there are one million – of whom 100,000 die every year from its complications. The annual cost of diagnosing and caring for Alzheimer's patients in the US was estimated to be $80 billion in 1991.

First diagnosed by the German neurologist Alois Alzheimer in 1906, the disease begins as recollections of recent events evaporate and feelings of time and place become disoriented. Then, insidiously, the power of language is demolished, memory is eradicated, and finally the victim's ability to eat, dress and control bodily functions is wiped out. Without constant supervision, victims will wander in front of traffic, leave on gas, or burn themselves. Essentially, the individual is annihilated in all senses except the physical. For the victim, the outcome is tragic. For his or her loved ones, left to cope with the shell of the person they once held dear, the progress of the disease is equally gruesome and catastrophic. 'The body remains as a constant reminder of a person who was once loved and indeed still is,' says President of the Alzheimer's Disease Society of Great Britain, Dr Jonathan Miller. 'It is a hideous memento of what you have lost.' Eventually victims do die, usually from complications of their immobility, such as bronchial pneumonia, though not before they have inflicted a crushing burden on their closest relatives.

Most Alzheimer's cases, which begin typically when patients are in their sixties and seventies, arise spontaneously, with no previous history being recorded in their families, though there is a second, smaller category in which intellectual impairment begins earlier, in middle age, and which does follow through lineages in some cases. In both varieties, however, biopsies reveal two very striking features. They show that the brain becomes filled with a protein known as amyloid beta protein, which is found in ugly clumps, called plaques, wrapped round nerve cells. In addition, tangles of filaments extend inside those nerve cells, effectively clogging them, blocking their functioning and eventually killing them. The greater the number of tangles and plaques, the worse are patients' symptoms. But what causes these plaque and tangle build-ups, researchers wanted to know. All sorts of explanations were proffered – from aluminium poisoning of drinking water (the metal is found in high concentrations in the cores of plaques) to breakdowns in neural transmitter circuitry.

However, as in other cases, it was the disease's inherited form that provided the decisive breakthrough, though as we shall see the deductive process was not without its false leads and blind alleys. One leading scientist was even forced to recant his views about the disease in a manner worthy of St Paul on the road to Damascus.

To crack the problem of Alzheimer's, geneticists needed to find a link between a patient's molecular biology and the appearance of plaques in the brain. It seemed like a classic problem for 'reverse genetics', a process that we have seen can be highly effective but also extremely laborious. However, in the case of Alzheimer's there was an extra clue that was to be of decisive importance. Children with Down's syndrome who survive to adulthood have brains that degenerate in much the same way as those with Alzheimer's. They even have similar plaques. And as Down's is caused by children being born with an extra section of chromosome 21, this strongly suggested an ideal starting point for research.

So Jim Gusella, of the Massachusetts General Hospital (who made the first linkage of the Huntington's chorea gene to chromosome 4), Peter Hislop and

Rudolph Tanzi, of Harvard Medical School, and John Hardy, then of St Mary's Medical School, London, and later of the University of South Florida, and several others began a close examination of chromosome 21. Their early results proved encouragingly fruitful. In 1987 several teams discovered a protein which naturally decays into amyloid beta protein in the brain. Critically the gene for this precursor protein is found on chromosome 21. As Professor Selkoe puts it: 'The discovery gave rise to a kind of global "aha!" experience in the Alzheimer field.' Scientists were excited because the link suggested a fairly simple mechanism: victims were inheriting a gene that makes a mutated form of amyloid precursor. In the brain, this decays leaving behind excessive amounts of amyloid – with devastating consequences for the nerve cells.

Unfortunately, linkage studies failed to substantiate this agreeably uncomplicated theory. In some cases, markers which aligned with the precursor gene on chromosome 21 were found to track through families along with the disease. Inheritance of that piece of DNA from chromosome 21 was the critical feature that seemed to determine which relatives would get Alzheimer's. However, researchers in Seattle found that in other families these markers just as assuredly did not track the disease. Their work indicated that a gene on a different chromosome must be involved. 'Something else is implicated. The amyloid gene is not the guilty party,' announced Dr Hardy.

These were to prove to be rash words, for a year later, in 1990, researchers showed that an amyloid precursor gene mutation was the cause of a disease called Dutch congophilic angiopathy, which bears a striking resemblances to Alzheimer's. The evidence was beginning to build up again. So Hardy's team re-examined their data and discovered, in one large British family, that inheritance of a version of the precursor gene was sufficient to cause Alzheimer's. Since then, two other types of precursor mutation have also been discovered – which explains the simple graffito on Dr Hardy's door: 'On this spot, John Hardy discovered amyloid was the gene.'

'The mistake we made was in thinking that genetic Alzheimer's was like cystic fibrosis, that all victims have mutations in the same gene,' says Dr Hardy. 'That meant that when we came across families with no connections with the amyloid gene, we assumed something else must be responsible for all cases.'

Behind the apparently conflicting evidence, scientists had found a fairly simple state of affairs. For a start they had shown that the symptoms of Alzheimer's disease begin when amyloid plaques build up in the cortex. It is these plaques that lead to the creation of tangles inside neurones (nerve cells) and eventually cause their deaths. The crucial point is that amyloid builds up in the brain for several reasons, some environmental and some genetic. What misled researchers was their expectation that there would only be one, purely genetic cause. It now appears there are several. The only one that has been uncovered so far concerns the gene which makes the chemical called the beta amyloid precursor protein.

Amyloid precursor protein is a large molecule of about 700 amino acids which has an important but yet unknown function in the brain. Once it has carried out its designated cortical task, the protein is naturally degraded by enzymes, the chemical foragers of the body, and amyloid – a small molecule of forty amino acids – is left over as indigestible refuse. And as we now know, it is the build-up of large amounts of this neurological detritus that causes the symptoms of Alzheimer's disease.

What Dr Hardy and his team found was a mutated gene which makes a version of precursor protein that leaves behind far greater deposits of amyloid than normal. Inheriting this defect, found on chromosome 21, leads inexorably to Alzheimer's. (This discovery also explains why Down's syndrome victims get Alzheimer's. With an extra copy of chromosome 21, they have an extra copy of the gene that makes the precursor protein. This generates more precursor protein and therefore more amyloid detritus.)

In other inherited cases, however, families have mutations in different, as yet unknown, genes – probably those which make the enzymes which normally break down the amyloid precursor protein. Again amyloid builds up, but for a completely different genetic reason, one that is unconnected with chromosome 21. This explains why the Seattle team found no linkage with their families.

And of course there are other, non-genetic reasons for amyloid to build up. One environmental trigger may involve high levels of aluminium in water supplies. Another clearly implicates injuries to the head, for both boxers and individuals who have suffered cranial damage in car and other accidents show disproportionate tendencies to succumb to Alzheimer's later in life. Severe cranial shock appears to set in motion an unknown process that slowly filters down the pathological pathways of the brain until amyloid builds up and the conscious mind is destroyed.

'Boxing and the amyloid precursor gene lie at opposite ends of the Alzheimer's spectrum,' says John Hardy. 'The former cause is purely environmental in nature, the latter is purely genetic. In between lie the vast majority of causes, which are both environmental and genetic in differing proportions.'

In fact Alzheimer's behaves in a manner strikingly reminiscent of heart disease. In coronary ailments, cholesterol builds up in the blood, eventually triggering a heart attack. In Alzheimer's, amyloid accumulates in the brain, eventually leading to nerve cell damage and symptoms of dementia.

Nor do the parallels stop there. In both conditions, there are purely genetic forms – arteriosclerosis and the amyloid precursor mutation. There are also purely environmental causes – eating fatty foods or drinking large amounts of aluminium-rich water. In the case of Alzheimer's, however, it was the insights offered by studying the disease's genetics which revealed the one, crucial common denominator in its development – the building up of amyloid in the brain. And that has been of inestimable importance.

In addition, other surprising similarities between Alzheimer's and heart disease have been uncovered. In 1993, scientists at Duke University discovered that a protein known as apolipoprotein E (ApoE), which was already thought to be a cause of heart disease, was coded for by a gene on chromosome 19. They also found that it came in three distinct variants: E2, E3 and E4. Crucially the Duke team discovered that the last version, E4, is found in high frequencies among Alzheimer's patients.

It was revealed that of those with one copy of the ApoE4 gene, 45 per cent have late-onset Alzheimer's disease by the age of seventy-five, while among those with two copies, the proportion rises to 90 per cent. The average age of onset reveals a similar pattern: 84 in the absence of the E4 gene, 75 if it is found on one chromosome, and 68 if it is found on both.

Such a discovery therefore makes it possible to identify people at very high risk of getting Alzheimer's later in life, even though the ApoE's function remains a mystery. But who will want to know if they are going to succumb to this mind-destroying disease in later-life? Very few, for no treatments for Alzheimer's disease have yet been created. The problem is one of the most awkward of those associated with the creation of genetic screening tests.

However, such research also brings forward the day when we might think of effectively countering Alzheimer's symptoms. For example, scientists soon hope to be able to create mouse models by inserting amyloid genes in mice, effectively giving them Alzheimer's. Then it should become possible to test drugs that might slow down the disease's symptoms or even halt the ravages of nerve cell death.

In short, the Alzheimer's story is a powerful indication of the might of the new genetics and a telling sign that many other neurological ailments should soon succumb to its attentions, especially when the Human Genome Project is fully underway.

9

Probing the Past

Two hundred thousand years ago, give or take the odd millennium, a strange alteration occurred within the human brain. A few minuscule changes to some ancestral neurones and our species was sent 'ticking like a fat, gold watch'.

At that time, Homo sapiens was represented by a handful of hominids who were struggling to survive in the burning savanna of eastern Africa. Yet the few modifications that affected their mental hard-wiring were enough to transform these minor characters in the saga of evolution into masters of the planet. The artistic glories, technologies, architectures and countless other wonders, and terrors, which have subsequently been created by modern mankind are the direct consequences of that neurological quantum leap.

Today we take the wonders of modern life for granted, forgetting they are the handiwork of only a very clever type of ape – as has been shown by molecular biologists who have turned their clinically precise techniques upon the history of mankind's evolution. They have found that the genetic void, which was once supposed to separate humans from all other primates, now appears to be the merest wafer-thin crack in the path that Homo sapiens has taken to world domination. Just why we succeeded while other hominids faltered and died out is now a matter of considerable research and speculation. After all, when we find out what propelled Homo sapiens first past the finishing line, we will have uncovered the last brush-stroke that created humanity in its final, full image. We will have touched on the very quintessence of being human.

Nor is the story of mankind's fresh evolutionary triumph over his fellow hominids the only unexpected finding made by geneticists. Their work has confirmed that our species is young, homogenous and adaptive. They have also found clear signs that the unremitting hand of natural selection is still shaping Homo sapiens in subtle ways, providing insights into the history of humanity's recent movements round the globe. Today we are accumulating an avalanche of information about our past – from the birth of modern man right up to contemporary historical events.

In short, genes are being exploited, not just for their medical knowledge, but as emissaries from the past, as messengers that are giving us a fresh perspective on lives, relationships, families – even feuds – that were once rendered opaque by the passage of time. Just as crumbling bones have opened up one dimension in the study of human antiquity, and the analysis of languages has thrown up evidence for a second, so genetic research is now helping us to complete a full, three-dimensional image of mankind's past.

Let us start this story with the emergence of modern mankind. Until comparatively recently it was thought that Homo sapiens' origins were fairly ancient, dating back about one million years to when our archaic hominid ancestor, Homo erectus, was supposed to have emerged from Africa and spread round the world. Then, in all the planet's diverse corners, islands, remote highland areas and valleys, these early humans – upright, tool- and fire-using, but slightly smaller-brained than us – slowly evolved in separate and dissimilar ways to produce Eskimos, pygmies, Australian Aborigines and all the other manifestly diverse peoples that populate Earth today. The theory stated that modern humanity had no single home, and that its constituent races were divided by fundamental and deep-rooted differences.

Take the relatively large nose of the average European. This distinctive facial appurtenance has been passed down to us from Homo erectus through the evolutionary medium of Neanderthal man, the theory goes. Put simply, in Europe, Homo erectus evolved into Neanderthals (Homo sapiens neanderthalensis) which then evolved into modern mankind (Homo sapiens sapiens, to give modern humans their proper title, though simple Homo sapiens is enough). The big hooter of Western man (and woman) is therefore an anatomical endowment of this ancient lineage. Or take the strong cheekbone of the Australian Aborigine. The theory proposed is that this is a genetic gift from Java man, a variety of Homo erectus, who lived in Indonesia about 700,000 years ago.

In other words, it was thought that mankind's different branches reflect intense evolutionary divides. However, there are nagging flaws, discomfiting pieces of evidence that do not fit this neat picture – like the discoveries in the 1980s of fossils of modern humans in caves in Israel and South Africa, which were shown to be 100,000 years old. Such dates simply do not fit with any picture of creatures who were supposed to have evolved in Europe from Neanderthals. There the earliest remains of modern humans date from around 40,000 years ago. Yet the Israeli and South African finds show Homo sapiens already thriving elsewhere in the world! Instead, the two sets of fossils suggest a very different genesis for modern humans – as a species that evolved in an area that fell roughly between Israel and South Africa, perhaps in sub-Saharan Africa.

Nor were these the only clues to suggest scientists must reappraise our picture of the dawn of modern humans. For instance, as new blood types were discovered between the 1920s and 1960s, scientists began to realize that a comparison between frequencies of these types in a population could reveal how closely it was related to another population. It could show how long ago the two had separated from a common ancestor race. The greater the dissimilarity between blood patterns (and those of other proteins), the longer they must have been evolving separately. The greater the similarity, the closer must be their historical association. For instance, the high frequencies of the B blood group among European gypsies gave the first real clue that they were Indian in recent origin. Both sets of people have elevated frequencies of the B blood group – about 50 per cent, compared with Northern Europe, where the figure is only 10 per cent.

The importance of this genetic goldmine was first appreciated by Professor Luca Cavalli-Sforza of Stanford University. Using the newly developed technique of electrophoresis, he was able to differentiate between proteins and their underlying genes on a large scale and draw up trees and timetables that tracked the unfolding of man's racial diversification. The figures he obtained showed that the maximum separation of the major human racial groups could not have occurred much more than 40,000 years ago. It was another nail in the coffin for the notion that the major racial groups in Africa, Europe and Asia reflect evolutionary differences that began to manifest themselves a million or more years ago.

Then in 1987 Allan Wilson of the University of California, Berkeley, entered the fray. Wilson and his colleague Vincent Sarich had already provoked controversy by suggesting that humans and chimpanzees, far from being very distant biological kin, were closely related. Dates of ancient bone had put the evolutionary divergence of humans and chimps at about 20 million years ago when our common primate ancestors were thought to have split into two groups. One evolved into Homo sapiens, the other into two species of chimpanzee, Pan paniscus, the pygmy chimpanzee or bonobo, and Pan troglodytes, the common chimpanzee. However, when Wilson and Sarich studied human and chimp DNA, they found their genomes differed by only 1.6 per cent – making humans and chimps very close biological cousins. And if our genomes differ only slightly, then it followed that chimps and humans could not have been evolving separately for very long, argued the two scientists. They placed the divergence date at only four to eight million years ago. The idea was controversial at the time, though Wilson and Sarich's dates are now believed to be much closer to the mark than were the old divergence times.

After publishing this research Wilson decided to study contemporary human evolutionary history, and turned, not to the normal DNA that has formed so much of this book's content, but to mitochondrial DNA. Mitochondrial DNA is found not in cell nuclei, but inside other intra-cellular objects called organelles. These structures lie in the cytoplasm, outside the nucleus, and are the power packs for our cells. Mitochondria manufacture ATP (adenosine tri-phosphate), the chemical that is the basic fuel for running our biological processes. Without mitochondrial organelles, we would run out of energy. In turn, each of these tiny organic generators has its own supply of genetic blueprint – mitochondrial DNA.

Now, mitochondrial DNA has two special properties that excite geneticists. Firstly it mutates about ten times more quickly than nuclear DNA, so that evolutionary alterations in structure become apparent with greater rapidity. It behaves like a fast-ticking clock and is ideally suited for the identification of recent genetic changes.

The second important property of mitochondrial DNA is that it is inherited only through the maternal line. In other words, your mitochondrial genes are exactly the same as your mother's, and her mother's, and her mother's mother's, and so on back into the mists of humanity's past. A father's mitochondrial DNA is not passed on and always ends up on the cutting-room floor of our reproductive processes. Unlike nuclear DNA, it is not reshuffled with each generation, and would remain unchanged were it not for the accumulated effects of occasional mutations. Sometimes a C replaces a G, or an A is substituted for a T, for example, and this feature can be exploited to shine a torch on our dim, genetic past.

Wilson reasoned that it should be possible to exploit this feature to create an entire branching tree made up of samples of mitochondrial DNA taken from different races of the world. The greater the number of mitochondrial DNA mutations found between two populations, the longer ago must have been their evolutionary divergence, he argued. The greater the similarity, the more recent must have been that digression. In this way, a vast family network, a sort of chronological chart for mankind, could be drawn up.

Using specimens taken from placenta, Wilson and his research partner Rebecca Cann tested 147 people from various ethnic groups. They measured the divergence of one small variable piece of their mitochondrial DNA and created a giant tree which linked all the various races of the world in a global genealogical chart.

The study produced three conclusions. Firstly, it revealed that very few mutational differences exist between the mitochondrial DNA of human beings, be they Eskimos, pygmies or Australian Aborigines – providing further support for the idea that we all had a very recent common ancestor about 200,000 years ago. Secondly, it created a tree with two main branches, one consisting solely of Africans, the other with some Africans and everyone else in the world. Cann and her colleagues reasoned that the simplest explanation for this tree shape was that our first common ancestor lived in Africa.

The third finding of the study was the discovery that Africans had slightly more mitochondrial DNA mutations among themselves compared to those within other races, implying they are a little older than other races, a notion which also supports the idea that mankind arose in Africa and, according to their data, very recently as well – about 200,000 years ago.

The paper, published in *Nature* in January 1987, generated headlines round the world, for Wilson and his colleagues had provided the most sensational evidence yet assembled in the mounting onslaught staged by modern genetics on the idea that our racial differences are fairly deep. Like those other research groups, whose blood and protein studies had suggested that underneath humanity's patina of superficial differences we are very alike indeed, Wilson and his colleagues were repainting our perception of our own past. It was the most striking addition to the idea that human similarity stems from our newness, from the fact that only a couple of hundred millennia ago, mankind consisted of a gathering of perhaps only a few thousand individuals from which we can all trace our ancestry today. That is why humans appear to be much less genetically variable than chimpanzees – possibly by a factor of ten. It is because mankind has only recently evolved from one tight little group of ancestors. It is one of the ironies of our evolution – that our nearest cousins should be much more genetically variable but are consigned to a band across Central Africa, while we who are stunningly alike, for all our superficial differences, have conquered the globe.

This theory does not dispute the notion that Homo erectus spread round the world more than a million years ago and then evolved into Neanderthals, Java men and others. However, it does contradict the idea that we evolved from these peoples. Instead, Wilson and his team suggest that a mere 100,000 to 200,000 years ago a second wave of humans – upstart newcomers, Homo sapiens – poured out of Africa and spread round the world, superseding all other primitive humans in the process. All the ancient lineages that had derived from Homo erectus, from Siberia to Spain and from Scandinavia to New Guinea, were overcome, with not one vestige, not one single genetic trace remaining. All were replaced. No intermingling, no mixing of genes took place, the Berkeley researchers insist. Instead, a single highly homogenous race of hominids – us – were left masters of the world.

Some palaeontologists, who had spent decades carefully building up a picture of mankind's fairly ancient divisions, were not surprisingly a little upset. One fossil expert, Professor Milford Wolpoff of the University of Michigan, denounced the whole enterprise as the 'Killer African' theory because he said it implied that modern man must have wilfully wiped out all other hominids in an orgy of violence, so quick was his conquest of the globe. 'This rendering of modern population dispersals is a story of making war, not love and if true its implications are not pleasant,' he claims.

It is a criticism derided by many other palaeontologists, however. Wolpoff's Killer African remarks are 'not reasonable', says Richard Leakey, for example. And for their part, the 'Out of Africa' supporters say that humanity's more ancient lines simply failed to compete effectively for resources – food and shelter – and slowly died out, though sometimes taking tens of thousands of years to do so. There need have been no violence at all, they state; an idea that is supported by Leakey.

But the theory has weaknesses, particularly as Wilson pushed its implications right to the limit, arguing that his mitochondrial tree could be traced back, not just to one race of hominids, Homo sapiens, but right back to one single woman who lived about 200,000 years ago in Africa. Newspapers predictably turned this idea into an image of a single mother who gave birth to the entire human race; a voluptuous, fecund matron who strode the savannas of east Africa nurturing a nascent race of 'superchildren'. Predictably, this mitochondrial matriarch was christened 'African Eve'.

The image of such a woman is a dramatic and powerful one, and does serve to encapsulate the idea of how small was the group from which geneticists say we so recently evolved. But the vision also has its limitations. There must have been thousands if not tens of thousands of women belonging to Homo sapiens at that time. Our roots go back to all of these women, and to their male partners. Our nuclear genes – which essentially delineate all our characteristics, as we have already seen – are a mosaic of contributions from a myriad of these ancestors. We appear to get our mitochondrial genes from only one woman, but that is most probably a consequence of the way that mitochondrial DNA is inherited.

Think of it as the female equivalent of passing on family surnames. When women marry, they usually lose their surname and assume their husband's. Now if a man has two children, there is a 25 per cent chance both will be daughters. When they marry, they too will change their name, and his surname will disappear. After 20 generations, 90 per cent of surnames will vanish this way, and within 10,000 generations (which would take us from the time of African Eve to the present day), there would only be one left.

Looking at the vast, single-named clan, one might assume that it bore a disproportionately high level of its originators' genes. In fact, it would contain a fairly complete mix of all human genes. In the same way, the mitochondrial DNA of one woman may dominate now, but this does not mean that our nuclear DNA, in which resides our true genetic diversity, does not have a rich variety of antecedents. Of course, such an analogy takes no account of people's habits of giving themselves new names, or more recently for not getting married at all and therefore not changing their names. However, it does show that we should take care in interpreting the implications of Wilson's rather provocative ideas.

And there were other criticisms. For a start, of the 147 individuals who had been used to supply the raw data, 98 had been found in American hospitals. And in particular, of the twenty 'Africans', only two were actually born there. The other eighteen were American blacks, though they were classified as Africans for the study. Given that so much had been made of its African results, the failure to sample directly from people who actually lived on the continent seems a little remiss.

The Berkeley researchers say this geographical inexactitude does not matter. Male American blacks, until very recently, did not produce children with white women. Inter-racial breeding was almost exclusively between black women and white men, the scientists argue. The African origins of the resulting offspring's mitochondrial DNA (from their mothers) would therefore have been preserved.

Nevertheless, changes in methodology were introduced in response to these criticisms, and in 1991 a second major paper was published shortly before Wilson's death from leukaemia. It employed more detailed analyses of mitochondrial DNA, and was based on a more reliable ethnic mix as a source of samples. Again it produced a branching tree which places mankind's birthplace firmly, and recently, in Africa.

Subsequent papers have questioned the mathematics used to create these branching trees, though not all the critics responsible have suggested these flaws invalidate the 'Out of Africa' theory. 'We looked at the data and found that three different trees could actually be produced from the mitochondrial patterns, not one,' says Dr Maryellen Ruvolo of Harvard University. 'Two have their roots in Africa but one has non-African roots. This means we cannot use the shape of the tree to infer our geographical origins, at least not using this part of the genome. Certainly, more work will have to be done to clarify the issue. However, all the rest of the mitochondrial data is entirely consistent with a recent origin in Africa, which is most likely the true homeland of modern mankind.'

Further support for the Out of Africa hypothesis subsequently arrived when Professor Cavalli-Sforza published the final results of his massive study of the structures of racial proteins and enzymes whose variations reflect underlying variations in nuclear DNA. This produced a similar branching tree for the human race, one that again suggests a recent African origin for Homo sapiens. 'Our work does not prove the Out of Africa model,' says one of the study's principal researchers, Professor Kenneth Kidd of Yale University. 'However, its results are entirely consistent with it, though there may well have been some admixture, some mingling of the genes that has not yet been picked up by researchers. It is hard to believe there was not some interbreeding between old and new humans.'

Intriguingly, Cavalli-Sforza and other researchers also looked at the languages of the subjects they were sampling, and created a tree of relatedness between the various tongues. The branching network produced this way was remarkably similar to that of their DNA tree. The history of our genes and our words go hand-in-hand, it seems.

Nor does backing for the Out of Africa hypothesis come only from geneticists and linguistics. Many leading palaeontologists have swung behind the theory, including Dr Chris Stringer, of the Human Origins Group at the Natural History Museum in London. His own research had led him to conclude that modern humans must have evolved separately and recently in Africa several years before Wilson's mitochondrial DNA study was published in 1987.

The fossils which Stringer finds especially relevant came from sites in both Africa and Israel. The latter include caves at Jebel Qafzeh near Nazareth, and Kebara on Mount Carmel. At the former site, archaeologists used the very latest techniques – thermoluminescence and electron-spin resonance – to date remains of Homo sapiens, and found them to be about 100,000 years old. Yet nearby, at Kebara, Neanderthal remains were dated at 60,000 years old. Now modern humans, according to the old theory, were supposed to have evolved from Neanderthals, yet they appear to have preceded them by at least 40,000 years! They may even have lived near each other with virtually no mixing, like two well-defined and separate species. 'Modern humans and Neanderthals seem to be distinct lines,' says Stringer. 'Both must have diverged from a common ancestor about 200,000 years ago – the Neanderthals evolving in Europe and moderns in Africa.'

Not everyone agrees with his interpretation, as Richard Leakey points out. Palaeontologists are divided equally over the issue, and many resent the interference of geneticists. 'As a fossil person myself I can sympathise with that sentiment,' says Leakey. 'When you look at fossils, you can see the anatomy, you can feel the morphology, and if you have the right eye for it, you can strive to identify evolutionary relationships. Fossils, after all, are the tangible remains of what actually happened in our history. But I have also learned of the potential power of genetic evidence. Any anthropologist who chooses to ignore it or characterises it as not real evidence does so at his or her peril.' For their part, geneticists point out that bones are only useful as guides to the past because they reflect changes in genes.

The human mitochondrial genome was finally sequenced in 1981. In medical terms, this sequencing has taught us a great deal about diseases, such as muscular illnesses, that are associated with mitochondrial impairments. However, the real revolution has been a cultural one. We have overturned previous ideas about our own origins – thanks to that sequencing.

And yet this work is really only a model for the full Human Genome Programme which, in comparison, is going to increase our knowledge about the roots of our different cultures, our heritages, and our relationships with each other, by several orders of magnitude. 'This is going to be the greatest archaeological excavation in history,' says Dr Svante Paabo of Munich University, a view shared by Stephen Jay Gould, the Harvard biologist. 'The reconstruction of the human family tree – its branching order, its timing, and its geography – may be within our grasp. Since this tree is the basic datum of history, hardly anything in intellectual life could be more important,' says Gould.

And one of the prime targets for this onslaught will be the Y chromosome, which might tell us, not about African Eve, but about her Adam. By studying a variable section of the Y chromosome it should be possible to create a branching tree for humanity's male line. (Just as mitochondrial DNA is inherited only from women, the Y chromosome can be inherited only from men.) And who knows what such a pedigree might show? It could, for example, prove to be more ancient than the female one, suggesting that Homo erectus males, or Neanderthal men, might have mated with some modern females, though their females may not have consorted with modern males.

Unfortunately no one has yet found such a conveniently variable region on the Y chromosome. No matter where they look, scientists have found that the Y chromosome, regardless of race or creed, has a molecular configuration that hardly varies at all. It does not matter whether the sample is Japanese, or comes from a Bushman, or from a South American Indian, its composition produces extraordinarily few distinctive characteristics and certainly none that can distinguish human populations from each other, though researchers are confident this position will soon improve. As Dr Ruvolo puts it: 'At present, this work seems to show – scientifically – that men are all the same.'

In the meantime, a veritable barrage of issues are gathering, waiting to be resolved when labs do make further developments with nuclear DNA. And of these, one particularly burning question just begs to be answered. Exactly what evolutionary advantage did Homo sapiens have over his hominid competitors, and in particular over our nearest evolutionary brothers and sisters, the Neanderthals? What genetic gifts made Homo sapiens so special and allowed us to inherit the Earth, while other hominids conspicuously failed?

Whatever these genetic endowments were, they could not have amounted to much in quantity of DNA change, for it is clear that the boundary between ourselves and Neanderthals must have been very slender indeed, and could only have been represented by changes in just a handful of attributes. To appreciate this last point, let us look again at our picture of anatomically modern humans emerging from Africa about 100,000 years ago. They must have come into contact very quickly with Neanderthals – and found them to be a large-brained, sturdy, sophisticated set of humans who buried their dead, and made stone tools. In other words, these were not ignorant cavemen – at least no more so than were Homo sapiens at that time.

In fact Neanderthals were a lot like us, though their features – deep eyes, prominent eyebrows, protruding teeth and jaws, and sloping foreheads – would have made them distinctive to our eyes. Underneath their skins, however, they must have been very similar to us, as a casual calculation indicates. Chimps and humans diverged six million years ago and our genomes differ by about 1.6 per cent, according to DNA hybridization studies. However, the divergence period between Neanderthals and modern humans was probably only about 200,000 years, which suggests we differed – very roughly – by only about 0.06 per cent of our genome. And that slender gap must account for our success and their failure.

The fact that our world domination may have depended on such a slight genetic difference indicates that our Great Leap Forward was the 'nearest run thing', an idea that is supported by fossil discoveries, like the remains of moderns and Neanderthals found in the Israeli caves mentioned previously. These show that the species overlapped for at least 40,000 years in the Middle East and for about 10,000 years in Europe. There both sets of humans spent their lives hunting, gathering fruit and berries, making stone tools, burying their dead and carrying out simple religious ceremonies (at least in the case of Homo sapiens). Today there is absolutely no apparent difference, no glimmer or hint left to explain why one would succeed and one would fail.

Then about 45,000 years ago some mystical quirk of behaviour seems to have been added to the contemporary human anatomy, a twist that produced innovative, fully modern people who proceeded to spread westward into Europe quickly supplanting Neanderthals.

What was that quirk? What slight advantage did we develop over this crucial period of our evolution, which the Neanderthals failed to produce? There is only one plausible answer, say many experts: it was the anatomical and neurological basis for spoken complex language. The missing ingredient involved fine modifications to the brain and to the human vocal tract – the larynx, tongue and associated muscles –which allowed modern man to communicate in more sophisticated ways and develop language that we would recognize as meaningful speech. Without it, we would probably still be stone-wielding hominids today.

'With language,' says Jared Diamond, professor of physiology at the University of California, Los Angeles, 'it takes only a few seconds to communicate the message, "Turn sharp right at the fourth tree and drive the male antelope towards the reddish boulder, where I'll hide to spear it." Without language, that message could be communicated only with difficulty, if at all.

'Without language, two proto-humans could not brainstorm together about how to devise a better tool, or about what a cave painting might mean. Without language, even one proto-human would have had difficulty thinking for himself or herself how to devise a better tool.'

In other words, it was the genetic capacity to speak a complex language that raised modern humans from the 40,000-year doldrums which we were then sharing with Neanderthals. It gave us the power to take over the world. This interpretation is shared by others, such as Professor Kidd of Yale's human genetics department. 'Our success must have had a lot to do with speech which is, after all, an enormously complex process. When we talk, more than 100,000 neuromuscular events are triggered every second, and the movements of more than 100 muscles have to be co-ordinated. The diaphragm, tongue, cheeks and jaw all have to be controlled. That whole process is extraordinarily difficult. For humans, this was a triumph of evolution and it set us apart from the rest of the animal kingdom.'

Armed with 'the perfect plainness of speech', humans would have been able to describe precisely where fruit and vegetables were growing, direct elaborate hunts and allow tribal elders to recount how famines had been conquered. Neanderthals may have had speech, but it would have been cruder and less effective than the sophisticated language of humans.

There is disagreement about the issue, however. Chris Stringer, for example, is a champion of a different attribute – memory. He believes those extra genes were devoted to the neural storage of information. 'There is no point in having language if you do not have the power to retain the complicated knowledge that you wish to pass on,' he says. 'With good memories you could have complex social relations. You could recall where you saw good hunting grounds the previous year and where you could find food supplies and vegetation.'

Then there is the idea, backed by American anthropologist Lewis Binford, that Homo sapiens possessed, but Neanderthals lacked, the genes that control the neurological power to plan in depth, an ability that allows us to foresee and plan for alterations in our circumstances and to map and exploit resources. Or then again it may just have been a simple matter of having a few extra genes which allowed us to live longer, so that there were more elders who could pass on the benefit of their wisdom, such as what had been done in their youth during serious drought.

Nor is the identity of mankind's mysterious advantage the only unresolved problem about our Earthly triumph over Neanderthals. There is also the question of the timing of its evolution. Did that vital neurological idiosyncrasy manifest itself 200,000 years ago, or 50,000 years ago? In other words, did we bring our slender mental advantage with us as we began our African exodus, and was it therefore so slight that its effects took 150,000 years to accumulate, before snowballing to provide Homo sapiens with eventual victory? Or did it occur later, and was it therefore more profound, and much speedier in its effects? Genetically, it is not possible yet to provide an answer, though archaeological evidence and anatomical and behavioural studies suggest the former case. 'There have been no significant changes in human anatomy for the past 100,000 years,' says Chris Stringer. 'That suggests, but does not prove, that the crucial evolutionary alterations which created Homo sapiens had already taken place by then. They just took a long time to take effect.'

Now these arguments have real fascination, but may seem academic, for none can be proved or disproved – at present. But when the Human Genome Project has been completed, and our DNA has been fully sequenced, we will have created a superb baseline which could be exploited by scientists to begin to answer such questions. Already archaeologists are attempting to isolate Neanderthal tissue and blood samples from old flints and bones, and once they do they might be able to compare their genomes with ours. Then we might discover which piece of genetic architecture gave our brains the kick-start that propelled us down the road to evolutionary supremacy. This proposition would certainly have seemed fantastic only a decade ago. Yet such has been the dramatic rise of the power of molecular biology that scientists are now seriously contemplating the notion.

Nor does the power of the molecular biologist stop at this basic analysis of mankind's first steps Out of Africa. With increasing exactitude, they are unravelling the strands of the great diaspora that followed this initial exodus, tracing humanity's progress round the globe right up to the present day. Cavalli-Sforza's study suggests a fundamental timetable for this colonization. Firstly, we moved from Africa to Asia about 100,000 years ago, and spread eastward until we reached New Guinea about 60,000 years ago. From there we moved to Australia (possibly by a land connection which may then have existed between these two 'islands'). A little later, having conquered the east, mankind turned westward and poured into Europe, finally extinguishing Neanderthals there.

Mankind's first movement into America is more difficult to establish, however. It must have occurred at some time when the Bering Strait was dry and the climate was mild enough to permit passage by land. The first confirmed settlement in America has been dated, in Alaska, at only 15,000 years ago. However, some archaeologists believe several sites in South America may date from 35,000 years ago. Intriguingly, the genetic data gathered by Cavalli-Sforza and his team support these earlier dates, for analyses of native American blood and proteins indicate that these are divergent enough in structure to suggest that the continent must have been settled at least thirty millennia ago. In fact, recent studies suggest there may have been three different waves of immigration from Asia.

At some point in the midst of these peregrinations, a few men and women changed in a way that was rudimentary and superficial but which has been the source of some of the greatest misunderstandings among humans: they evolved a different coloured skin. In other words, at some time it became an advantage for some people to evolve light skin, changing pigmentation from the original dark-skin prototypes of modern humanity.

A light skin provides a benefit where sunlight is poor, since it enhances the formation of vitamin D, a lack of which leads to rickets. Crippled with rickets, our early hunter-gatherer forebears would not have survived long. So it is probable that this genetically determined attribute was initially selected for the extreme north of Europe where sunlight is poor. After this the genes for light skin spread, partly because it provided an advantage in murky climes but it may also have become valued for its appearance. Those with the gene – individuals who are seen to be striking and unusual in looks – may have been more desired and therefore may have mated more often than others. This would have enhanced the spread of the gene.

Since then, the world has been convulsed with continual waves of migrations and invasions, and many have since given up their secrets, through studies of fossils and now through genetic research. In general, both approaches, the Neanderthal issue notwithstanding, have confirmed each other's findings. Yet there remains a crucial difference between them. Standard archaeology is a process for dealing with the dead and the fossilized. The startling feature about its genetic equivalent is that it deals very much with the living, with the very DNA that controls the behaviour of each of our cells. In other words, instead of using the fossilized past as a window on the

NUMBERS REFER TO ESTIMATED ANTIQUITY
OF FIRST SETTLEMENT, IN YEARS.

15-35,000

35,000

100,000

60,000?

>40,000
(50-60,000?)

Out of Africa. Mankind's routes to the global conquest that we initiated only 100,000 years ago. The light lines represent a phylogenetic tree of the major human population groups superimposed on the postulated patters of migration out of Africa.

present, genetics allows us to use the living present to understand the past.

It is an approach that brings particular advantages, as Allan Wilson and Rebecca Cann point out in a *Scientific American* feature. 'Living genes must have ancestors, whereas dead fossils may not have descendants. Molecular biologists therefore know the genes they are examining must have been passed through lineages that survived to the present; palaeontologists cannot be sure that the fossils they examine do not lead down an evolutionary blind alley.'

An example of the power of this living archaeology is provided by looking at a gene map showing rhesus negative distribution in Europe and Asia, as Cavalli-Sforza did. With that we see a clear gradient. At Europe's western edge more than 25 per cent of the population is rhesus negative. As you head east, its frequency drops off among the local populations, declining steadily. By the time you reach the Middle East, only 5 per cent of the population is rhesus negative.

Now this decline in negativity has a simple origin. Western European blood (i.e. blood with high frequencies of rhesus negative groups) was probably the normal variety in Europe 10,000 years ago. Then from the east came the farmers. Out of the Balkans and from the Middle East, the first neolithic practitioners of the science of agriculture moved into virgin hunting ground in search of land on which to plant and to graze. And in the wake of their economic expansion, the farmers brought their culture, their languages – and their genes. And this last feature was conspicuous for its relative lack of the rhesus negative blood gene. Crucially, these rhesus studies have been supported by other gene surveys, such as those for HLA type distribution.

Farming advanced at a rate of about one mile a year, with new settlements being founded at the edge of freshly cultivated areas. The newcomers were no doubt resented by the indigenous populations, but it is equally clear that they mixed, intermarried and bred. And farming would have supported considerably more people than previous practices, which would have fuelled the migrational advance.

Eastern European people were obviously the first to come into contact with these farming *arrivistes* and will have absorbed their genes for longer. We should expect rhesus negative blood to be relatively rare among them and that, indeed, is what we find. In the West, we find the opposite is true. The forebears of the Basques lived right at the end of this migratory path, for example. Contact, if any, came late so that locals underwent the least genetic mixing with the farmers. Indeed, to judge from Basque blood – and several other sets of their genes and also language – it seems as if their ancestors resisted the incoming farming technology altogether and may therefore be close representatives of the original hunting people of Europe.

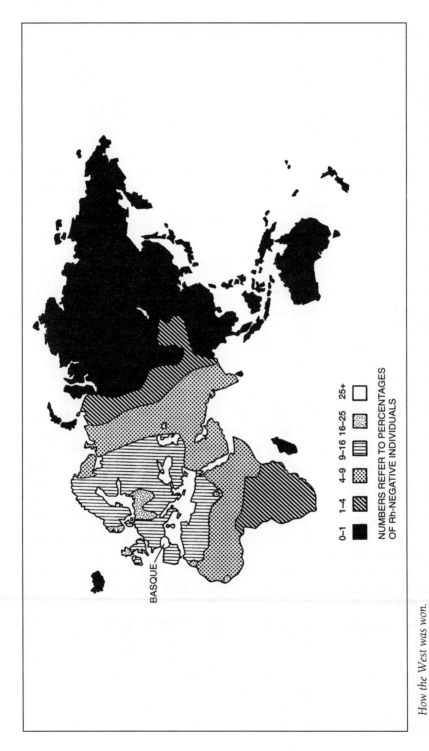

**NUMBERS REFER TO PERCENTAGES
OF Rh-NEGATIVE INDIVIDUALS**

0-1   1-4   4-9   9-16   16-25   25+

How the West was won.

The map of the frequency of the rhesus negative gene reveals a tell-tale pattern and shows that the spread of farming from the Middle East is still reflected in the blood of Europeans today. Those areas in the extreme West, particularly those in the Basque region, display the highest frequency of rhesus negative blood and probably contain people who are the closest representatives genetically – to the original hunter-gatherers of Europe.

The westward spread of farming across Europe was known about before modern genetic research was developed, of course, though there is something quite thrilling in realizing that the blood of these founding fathers and mothers of civilization can be detected in the veins of people today. And by looking at our past in this way we can see for the first time that farming moved not as a technological idea that was handed on, or was adopted by people who could see its obvious merits. Instead, it must have been physically brought in by migrating farmers, men and women who introduced the new technology and who carried it ever westwards. It is a singular way of looking at our past, one of many that are being opened by such analyses, some where no other technique is applicable.

Another example is provided by the discovery of high frequencies of the A blood group gene among the people of south-east Little England, that part of the Pembrokeshire coast in Wales which protrudes into the Irish Sea below the town of Pembroke. While the rest of the region has frequencies of the A blood group of less than 30 per cent, those of the people of south-east Little England are well above this figure. The peninsula sticks out noticeably in any A gene distribution map of Wales. The question is: how did this unusual configuration come about? Clearly some mixing of genes with immigrants took place some time in Pembrokeshire's past – but whose genes were responsible?

The answer can be found in early British history when the invading army of William the Conqueror established sovereignty in England and Wales in 1066. One of William's first acts was to bring over craftsmen from Flanders, an area that today still has a relatively high frequency of the A blood group among its residents. These workmen set up a colony in Little England. All the physical signs of that settlement have now disappeared, but in the blood of Pembrokeshire people today the tell-tale signs of their Norman past linger on.

Many other examples have been uncovered thanks to such gene surveys. For instance, studies of black Americans have shown that their gene frequencies differ from those of Africans, their immediate ancestors. Although black people were brought to America relatively recently, between the sixteenth and eighteenth centuries, a mere ten to five generations ago, there are already noticeable deviations between the genetic profiles of the two populations. It is now clear there has been a significant input of white genes into the black American genotype. Indeed, for the entire American black population, the percentage of white ancestry is estimated to be about 30 per cent. And given that marriage across racial barriers was effectively forbidden in the United States until a few years ago, the only realistic explanation conclusion is that a great many extramarital matings must have taken place between black and white people.

However, this gene geography has, until recently, been only an indirect science. It has measured genetic shifts, but only at the protein level. Underlying changes in DNA have been inferred through fluctuations in these protein and enzyme products. Now modern molecular biology is transforming that position, allowing scientists to examine genes directly; and it has sharpened the tools of the gene geographer by an order of magnitude. Much larger numbers of differences can be explored, producing much greater precision.

This honing of the biological archaeologist's tools has only just begun, but already encouraging results are emerging. One leading research group is based at the Institute of Molecular Medicine in Oxford where researchers have used precise DNA analysis to make some notable discoveries about the peoples of the Pacific.

Their work reads like a detective story, which began in 1978 when Don Bowden, the chief medical officer for Vanuatu, the Pacific archipelago once known as the New Hebrides, noticed unexpectedly large numbers of patients suffering from anaemia. Fellow doctors blamed hookworm, poor diets and other factors and advised treatments with iron supplements, an approach Dr Bowden found to be quite unsuccessful. So he contacted scientists at the Oxford Institute and sent samples of his patients' blood.

To everyone's considerable surprise, half Dr Bowden's donors were found to be carriers of the genetic blood disorder alpha thalassaemia, a condition that no one had previously realized affected Vanuatu. (Thalassaemia is an inherited, recessive failure to make haemoglobin, and in its carrier status it is linked, like sickle cell anaemia, to protection against malaria.) A bigger survey, of 2,500 people from islands ranging right across Melanesia, from New Guinea to New Caledonia, confirmed that the disease was widespread in the islands and found, as expected, a close correlation between prevalence of the alpha thalassaemia gene and the incidence of malaria. Both were frequent in the north-west of the region, in coastal New Guinea, and both tailed off in severity as you headed south and east to New Caledonia. It was some of the most conclusive evidence that a causal connection exists between malaria and thalassaemia.

But what sort of background did their thalassaemia natives have? In other words, had the condition arisen locally, or had it been imported by settlers? Careful study of chromosome types showed that the former was true. Their particular mutant gene was found only in chromosomes that were typically Melanesian and had not arisen in any other area.

The particular variant of alpha thalassaemia discovered by the researchers is known as 'the alpha 3.7 three deletion', terms which firstly tell you that 3.7 kilobases, or 3,700 bases, had been deleted from a carrier's alpha-

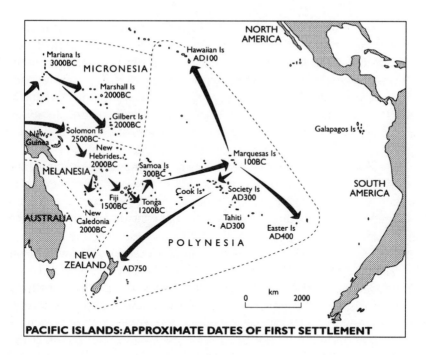

PACIFIC ISLANDS: APPROXIMATE DATES OF FIRST SETTLEMENT

haemoglobin gene. The whole gene complex is about 30 kilobases in length and the loss of the 3.7 section prevents it from making alpha-haemoglobin properly in bone marrow. The number three indicates that this is the third of three superficially similar mutations of this nature that have been discovered by scientists. It is also the rarest.

So far, so good. Researchers had traced Vanuatu's cases of anaemia to a very rare form of thalassaemia and linked its prevalence to the incidence of malaria. Then they began to look at other Pacific islands, particularly in Polynesia, thousands of miles to the east, and they began with Tahiti. To their considerable surprise, they found a quite high frequency of alpha thalassaemia among local people there as well, even though there is no malaria on the island, nor has there ever been. 'Thalassaemia, but no malaria, it seemed to throw our ideas about genetics out of the window,' recalls Dr John Clegg of Oxford's Institute of Molecular Medicine.

Then the researchers discovered that Tahiti's alpha thalassaemia was of a very specific sub-type – the rare 3.7 mutation discovered on Vanuatu. The conclusion seemed inescapable. The mutation had arisen in Vanuatu, or on an island close by, had been amplified as a protection against local malaria, then transported by ancient settlers out into the eastern, malaria-free Pacific – to be picked up 3,000 years later like a discarded genetic calling-card.

This discovery casts a revealing light on one of the most remarkable acts of colonization and seafaring in human history, when the forebears of modern Polynesians set out in tiny outrigger canoes loaded only with coconuts, yams, breadfruits, pigs and a great deal of skill and luck, to colonize the vastness of the Pacific. The venture began around 3000BC at the eastern edge of the Pacific, probably around Borneo, and ended four millennia later when Easter Island and New Zealand were settled – in AD400 and AD750 respectively.

This is the version supported by linguistics and archaeology and it states unequivocally that the Pacific was settled in an eastward direction from Asia to Easter Island. On the other hand, a small minority believe the ocean was conquered in a westerly direction. They say that Chile and Ecuador were the homelands of the people who set sail, westwards, to settle in the Marquesas, Tahiti and Easter Island, an idea first championed by Thor Heyerdahl who even sailed his raft *Kon-Tiki* from Ecuador to Polynesia to prove it.

But if Heyerdahl's theory is correct, it has to explain how the 3.7 version of alpha thalassaemia became a common feature of Polynesian people. It had to be imported into malaria-free Polynesia, after all, because otherwise its high prevalence cannot be explained. The common incidence of 3.7 alpha thalassaemia needs malaria to amplify it at one point in its history. (It is worth emphasizing again that, like sickle cell anaemia, thalassaemia carriers have some form of protection against malaria.) And if Heyerdahl's theory is correct, then it must have come from South America. Yet attempts to find the gene there have drawn a complete blank. 'The 3.7 mutation went from west to east, and not east to west, and the same goes for the settlers of the Pacific,' adds Dr Clegg.

However, there is more to be learned from this piece of genetic detective work than merely disproving the rather romantic ideas of Heyerdahl. It is known that two major waves of colonization were involved in the Pacific's conquest. Firstly Melanesia was settled by colonizers from Asia around 2000BC. Then 1,500 years later a second wave of people set off in their boats from Asia and headed eastwards, through the islands of Melanesia, which had been colonized by the first wave of settlers, and much further out into the Pacific – to Samoa, Tahiti, Easter Island and the rest of Polynesia. The question is, what sort of contact was there between the two waves? Did it involve only trade and the odd quarrel? Or was there considerable mingling and interbreeding? Linguistic and archaeological evidence suggest the latter notion is correct. And now thanks to molecular biology there is additional genetic proof – for there, like a biological flag, the 3.7 mutation shows that a Melanesian gene is now widespread in Polynesia. There had been considerable mingling and interbreeding. Scientists have been able to detect a precise connection between two peoples in a way that was previously impossible.

'We now know there was intermingling between the two races, for there is a significant input of Melanesian genes into Polynesian genes,' states Dr Rosalind Harding of the Oxford Institute of Molecular Medicine. 'That is the true significance of our 3.7 deletion discovery.'

The story of Vanuatu and the thalassaemia which evolved there is a particularly striking one. Apart from demonstrating the power of modern biological techniques, it also highlights another aspect of the evolution of mankind – its continuity. We should never imagine that the process of natural selection will ever cease, for all the stabilizing effects that have been introduced by modern society. Our species evolves continuously and we can see this process in operation with conditions like thalassaemia.

However, at an individual level, thalassaemia's effects are disastrous. A man and a woman – usually of a Mediterranean or Asian origin – can meet, fall in love and begin a family, unaware that both are symptomless carriers of this recessive ailment. Then they start to produce children suffering from the severe anaemia that is the disease's principal symptom. Only recently, with the advent of modern drugs, has it been possible to keep such offspring alive beyond adolescence.

In reality, thalassaemia is a Faustian bargain with nature. Mankind has gained in some ways out of the arrangement for, as we have already seen in Chapter Three, carriers of a sickle cell anaemia gene are better protected against the most common and dangerous form of malaria, caused by *Plasmodium falciparum*. This is also true for thalassaemia. However, as numbers of carriers increase, so do the chances that two will meet and have children. Some of these will inherit two thalassaemia genes, and they and their parents will be the losers in this pact.

In effect, thalassaemia and its 'cousin', the inherited ailment, sickle cell anaemia, is an ad hoc, imperfect response to extreme provocation by nature: malaria. The fact that there are several dozen different varieties of thalassaemia gene suggests that given long enough, a version will appear – by mutation – in a population. And if that population is affected by malaria, then that thalassaemia gene will spread, by conferring a small amount of longevity to those who carry it, though just how the gene protects against malaria we do not yet know. Then as the thalassaemia gene spreads through a population, more and more carriers will meet, and produce more and more severely anaemic children.

The occurrence of a thalassaemia gene can therefore be treated as a signpost to the past, for if it is prevalent among a group of people then they, at some time in their history, must have been exposed to malaria, as with the settlers of Tahiti.

This genetic message from our forebears is therefore saying something quite special about the environment from which we originated. Nor is thalassaemia the only illustration. There are the HLA genes which have evolved in response to past infections and which were discussed in Chapter Seven. The question is, are there any more such genes that have arisen as protective responses? And can we use them to throw light upon the pasts of other peoples?

The honest answer to both questions is that we simply do not know at present. However, there are some intriguing suggestions, such as the gene for Tay Sachs disease. This recessively inherited illness causes mental retardation and blindness and is common among Ashkenazi Jews – those of German and East European origin. The birth rate among them is about one in 2,500. Among the general population of the West, the birth rate is only one in 360,000, while among Sephardic Jews – those of Spanish or North African origin – it is one in 100,000.

Some scientists believe that carriers of the Tay Sachs disease gene are more resistant to tuberculosis, an illness that was rife in the overcrowded ghettos of Eastern Europe from which Ashkenazis probably originated. Just as the thalassaemia gene spread because of its power to stave off the worst ravages of malaria, so the Tay Sachs gene may well have arisen among Ashkenazis and spread through them, endowing a limited protection against tuberculosis.

And there is a great deal of circumstantial evidence to support the idea. For example, in the days of the ghettos Ashkenazi Jews appear to have suffered from a lower frequency of tuberculosis deaths than non-Jewish populations from the same area. In addition, the highest frequency of the Tay Sachs gene in Ashkenazi Jews was found in Austria-Hungary, which had the highest incidence of tuberculosis.

It is a fascinating footnote to one of history's less attractive periods, though the evidence for a link between the two diseases remains speculative at present. However, such protective links are only one of the mechanisms that geneticists use to account for the genetic pot-pourri of modern humanity. In particular, there is another very powerful force shaping mankind. The following story, provided by Jared Diamond, gives a telling demonstration.

Baby Pierre was born in Canada on 7 March 1964, at a healthy six and three-quarter pounds. However, from the very start he ate poorly and had to be hospitalized by the time he was six months old. He had gained only half a pound, his muscles were weak, and he vomited periodically. Strangest of all, his urine always smelled of rotten cabbage, and the odour permeated his body and clothes. In hospital, little Pierre's condition continued to deteriorate and on 30 November that year he died.

What really puzzled physicians was the number of similar cases they were uncovering at that time; most of them from the remote Chicoutimi region north of Quebec city. There some parents lost three or four children to this mysterious killer and reported similar deaths in previous generations.

Eventually the culprit was tracked down – hereditary tyrosinemia. The killer of baby Pierre was a defective gene which prevents the body from making a liver enzyme needed to break down the amino acid tyrosine. Without this enzyme, tyrosine accumulates and damages both liver and kidneys (which explains the cabbage smell, from excreted amino acids). Throughout most of the world, hereditary tyrosinemia – a recessive condition – is rare (about 1 per 100,000 births). But in Chicoutimi, one in 685 newborns is affected and about one in fourteen of the local population is a carrier.

So how did this lethal concentration arise? The answer has been provided by some clever historical and genealogical detective work by geneticist Dr Claude Leberge, who has discovered that all the pedigrees of Chicoutimi's tyrosinemia can be traced back to just one couple, Louis Gagne and his wife Marie Michel, who emigrated to Quebec from France in the seventeenth century. Some of their children were the original settlers of Chicoutimi and along with their farm implements, household goods and cattle, they must have carried the tyrosinemia gene like a piece of biological baggage. Settled by only a handful of people, several of whom carried the Gagne curse, the local population became disproportionately encumbered with the tyrosinemia gene, which explains why Chicoutimi is so affected with the disease even today.

This phenomenon is known as the founder effect, a term coined by the evolutionary biologist Ernst Mayr. It occurs when a population becomes genetically distinct as a result of being founded by only a few individuals whose genes are a biased sample of the genes from the ancestral population from which they are drawn. There are many fascinating examples of the founder effect and its impact on human populations.

Porphyria is a good example. It was established in South Africa by a pair of Dutch immigrants, Gerrit Jansz and Ariaantje Jacobs, who married there in 1688. Today, about 30,000 of their South African descendants (about one in 300 of the white population there) carry the gene that they brought – a far higher incidence than in Holland. As a result, because anaesthesia, barbiturates and other drugs can be fatal to sufferers, many South African hospitals now routinely test every admitted patient for porphyria. Like the Gagne curse in Chicoutimi, the Janszes have left their genetic imprint on modern South Africa in a lasting way.

In fact, medical textbooks are full of stories of genetic conditions, like porphyria or hereditary tyrosinemia, which affect particular localities. For instance, Huntington's chorea, the fatal inherited neurological disease discussed in Chapter Five, is disproportionately prevalent in one area of South Wales and can be traced back to one man, a mason from Devon, who came to the hamlet of Mynydd Islwyn in 1854, attracted by the rapid expansion in coal mining there. He carried the Huntington gene and passed it on to his descendants. Today more than 120 families in the area are affected by Huntington's chorea.

Then there are the Amish. Apart from their putative manic depression lineage, they are affected by an inherited condition which causes individuals to grow six fingers on each hand, and to be dwarfs. The source of this ailment has been traced to one of the few founders of the Amish community, a Mr Samuel King. Either Mr King or his wife happened to carry the gene for six-fingered dwarfism, which is now fairly widespread in the community.

Most fascinating of all, however, is the story of Martha's Vineyard, a small island off Cape Cod which is today a haven of holiday houses, but which was once the home of some of the world's most daring whalers. Thanks to some intriguing detective work by Harvard anthropologist Nora Ellen Groce we know that up until the turn of the last century Martha's Vineyard was affected by hereditary deafness. Indeed, so common was the condition that islanders considered it as a natural way of life. 'It was as if somebody had brown eyes and somebody had blue,' one islander told Groce. Deafness was no drawback to being a fisherman or farmer, the main occupations, so the condition was not viewed as a stigma. A large number of islanders – both those who were deaf and those who could hear – learned sign language. Many were, in effect, bilingual. As to the source of the deafness gene, it probably arrived with one of the forty-eight families who came to Martha's Vineyard in the sixteenth and seventeenth centuries, settlers who emigrated to avoid the religious persecution and poverty of their native Kent.

Now these examples of the founder effect may seem exotic, but only of peripheral importance in our unravelling of the saga of humanity's global conquest. After all, they involve ailments that are infrequent by normal standards, and affect populations that were founded only relatively recently. But we should not forget that half the land on our planet – i.e. North America, South America and Australia – was settled only a few millennia ago. The Americas were populated at least 15,000 years ago (or perhaps even longer ago, as we have seen) when bands of hunters first began crossing the Bering Strait and then spread southwards, ultimately reaching Argentina and Chile. Similarly, Australia was settled about 60,000 years ago by groups who had sailed from Indonesia. In both cases then, very large areas were conquered by small numbers of people whose genes were therefore elevated to new levels of frequency. And since then, sudden leaps in technological proficiency –

the development of guns, steel or ships, for example – have triggered sudden new waves of expansion and bouts of colonizations round the world. 'Almost any human population might show evidence of the founder effect, because the populations of three of the six inhabited continents were founded *de novo* 10,000 to 50,000 years ago, and much of the population of the remaining three continents was refounded in the last 2,000 to 8,000 years,' says Jared Diamond.

In this way, jumps or shifts in gene patterns have been created continuously since mankind first moved out of Africa. And in the due course of events deleterious genes, such as those for tyrosinemia or porphyria, would normally be eliminated by natural selection. Others would not be, such as some variations in blood group which may be due to the founder effect. Such neutral characteristics will persist to the present day, even if established long ago, for there is no evolutionary pressure to eradicate them. These are the legacies of the early founding fathers and mothers, men and women who have carried on the seemingly endless process of reconstituting mankind in their own image in different parts of the globe.

In general, however, the shifting nature of Earth's human gene map has been a large-scale business involving waves of migrants, such as those who brought farming to Europe, who have supplanted incumbent populations. The founder effect was more limited in its impact. We should also remember that these living undulations generally carried no new genes. The fundamental divide, then and today, between two populations is measured merely in differences in gene frequencies. All human races possess the same basic genetic stockpile. Differences occur only in terms of how often a given gene appears, such as the rhesus negative gene among Europeans and the first farmers from the Middle East. Armed with this knowledge, we can now detect past actions of long dead people. At the same time, this basic feature of human genetics also demonstrates the profound homogeneity of the human race.

Living human beings are therefore an unparalleled source of information about the past, and with the development of new genetic methods we should soon be able to provide even more exciting insights into our history. The trouble is that the basic source of raw data is under threat, for the world's indigenous peoples – the Bushmen of South Africa, the Hill People of New Guinea, Yanomami Indians of the Amazon rain forest and many others – are disappearing. Each of these populations has been isolated for millennia and offers a unique perspective on humanity's peregrinations. They offer modern science 'a window on the past', as Professor Kidd puts it. But disease, loss of hunting grounds and assimilation with Westerners now threaten their genetic integrity. Indeed for some people, such as the peoples of Basra, in

Iraq, who are believed to be the descendants of the ancient Sumerians, it may be too late.

One of the most important components of the Human Genome Project will therefore involve the setting up of a type of rescue genetic archaeology. This will consist of the sampling of all the world's threatened peoples before their unique stock is lost or mixed with Western genes; this is one aim of the Human Genome Diversity Project, which has been set up to sample blood and tissues from indigenous people all over the world in order to create a unique genetic collection of the different human races.

One of the project's many centres is Professor Kidd's laboratory at Yale University. There huge liquid nitrogen freezers are being filled with blood samples taken from races and tribes from around the globe. By the beginning of 1992, Kidd and his colleagues had collected more than 800 specimens obtained by anthropologists from the Baika pygmies of Central Africa, Cambodians, Basques, New Guineans, Samoans, Yemenite Jews, Ethiopian Jews, Malayans, Sardinians and a host of different ethnic populations. Many other laboratories have accumulated equally large genetic repositories. In addition, a related project is looking in greater detail at European populations.

Once a blood sample is sent to a centre, it is placed in a centrifuge and its B lymphocyte cells are removed. These are then infected with Epstein Barr virus, the causative agent of glandular fever. This particular virus triggers a process by which the B lymphocytes start to divide uncontrollably, the end result being an eternal supply of cells which, of course, contain the DNA of the blood donor. These cell lines become a permanent source of his or her genes and, repeated over and over again for all the world's different peoples, can be used to establish a living museum of human diversity, a repository of our genetic variability which can then be probed by the latest techniques of molecular biology. Creating this molecular Noah's Ark will be one of the most significant acts of the Human Genome Project. And in the next chapter we shall look at the most exciting of the techniques that will be used to probe our global gene bank, inventions that are revolutionizing not just research but also detective work, legal practice, immigration, and many other aspects of modern society. They reveal vividly how modern genetics is changing the day-to-day running of our lives.

10

Probing the Present

In the early hours of 20 December 1988, twenty-two-year-old Lorraine Benson was returning from an office Christmas party when she was attacked and dragged up a quiet alley as she walked home from Raynes Park railway station in south London. There she was assaulted, battered and strangled. The next day, her body was found on waste ground nearby.

Although witnesses reported hearing a woman's screams at the time of the murder, no one caught a glimpse of her killer. So the police began to comb the alley and waste ground for clues, amassing a cardboard boxful of grim detritus which was taken to the Metropolitan Police's forensic headquarters in Lambeth and handed to scientific investigator Julie Allard.

Among the macabre trophies was a man's bloodstained handkerchief which had been discovered in the alley a quarter of a mile from Lorraine's body. Was the blood Lorraine's and could the handkerchief have been dropped by the murderer, the police wanted to know?

These were good questions. Yet the forensic scientists at Lambeth would have been hard pushed to answer them only a few months earlier. 'Our laboratory had only just introduced the technique of genetic fingerprinting as one of our routine methodologies,' recalls Ms Allard. 'Fortunately it was ideal for providing answers to questions like those.'

Invented by Professor Alec Jeffreys of Leicester University, genetic finger-printing exploits those variable chunks of DNA that pepper our chromosomes. His breakthrough was to discover one type of DNA sequence that is repeated over and over again – but in a way that differs in numbers of reiterations from person to person. By using DNA hybridization techniques and radioactive labelling, these repetitions could be counted, he realized, and could be made to appear as a bar code that uniquely defines a person.

Alec Jeffreys.

Most uses of genetic fingerprinting have subsequently concentrated on testing blood and semen as a means to identify rapists or murderers, or simply to prove the paternity or maternity of a child. But Julie Allard wanted to know if the test could pinpoint the identity of a man from a very different secretion, in this case a crusty nasal stain that she had found on the handkerchief. 'There was no report of anyone ever having obtained a genetic fingerprint from nasal mucus at that time,' says Allard. 'But in theory it seemed obvious that it would be a good source of DNA. So I thought, why not go for it?'

Nasal mucus is produced when we sneeze and is made up of a mixture of bacteria which have been ejected from the body, and epithelial cells which have been sloughed from the respiratory glands during this violent exhalation. It is therefore a rich source of a person's DNA. So Allard ran a genetic fingerprint test on the mucus and obtained a clear print of an unknown individual.

Then Allard tested the blood. And again she obtained a clear set of bar code stripes, though this time the genetic fingerprint was identified. It was Lorraine's. The question was: had Lorraine bled over the murderer's own handkerchief, or over one that had been dropped there much earlier?

The answer was soon provided. On 2 February the next year, a young woman was the victim of an attempted rape at a house near Raynes Park station. Fingerprints left on a french window led police to John Dunne, a local thug who, although only nineteen years of age, had amassed a depressing string of convictions for robbery, attempted burglaries, rape and assault. Scientists made a genetic fingerprint of his blood, and got a perfect match with the one made from the nasal stain on the handkerchief.

But Dunne denied murdering Lorraine, claiming that he had lost his handkerchief in the locality several days before the murder. Someone else had killed her, he maintained. Detectives did not believe him but were temporarily stymied. Fortunately, the investigators were not finished. A clear imprint of teeth marks had been found on Lorraine's arm. So a cast of Dunne's teeth was taken – which produced marks identical to those on Lorraine. In his frenzy, he had bitten his victim. Then outlines on dirt on an abandoned car found in the alley near the murder scene were shown to match with a zip on Dunne's coat, and from one on Lorraine's. Allard was able to demonstrate that these two dusty palimpsests could have been created contemporaneously, which showed that Dunne and Lorraine had probably been beside the car at the same time. That was enough. Dunne confessed and on 23 October 1989 was sentenced to life imprisonment.

The case produced lead stories in most British newspapers, which is not surprising given the savagery of Dunne's crime. But because he pleaded guilty, only sketchy details of the scientific evidence against him were read out in court. The public never learned how Julie Allard's deft forensic footwork helped to trap a killer. In doing so, she had made use of an arsenal of different techniques which enabled her to link Dunne with his crime. And of these, the most decisive was clearly genetic fingerprinting. It was the telltale black bars of Dunne's DNA profile, provided by the handkerchief, that connected him unequivocally with the crime in the first place. The rest was inevitable.

Since then, genetic fingerprinting has established the guilt of hundreds of other rapists and killers, while also exonerating a great number of innocent people. Just as our genes are being used as couriers from our past in order to unravel human history, so sections of a person's genome can now be explored for clues about his or her own past actions. It is a development of considerable might, so let us look at the technique in a little more detail.

Discovered in 1984, genetic fingerprinting has an unusual pedigree; one that can be traced back to, of all things, a deep-freeze locker at the British Antarctic Survey's Cambridge headquarters. It was here, among frozen meat and tissue samples sent back by polar researchers, that Alec Jeffreys began experiments that would ultimately make headline news round the world. He was then investigating variations among genes; not those that produce abnormal products, like the one that causes sickle cell anaemia, but in harmless, neutral variants which still code for normal proteins. He was trying to improve methods for tracing genes through lineages, from children to parents, and to grandparents, and wanted to isolate DNA regions that varied distinctively between families, and which could then be employed as markers.

'We were investigating the genes that make myoglobin, a sibling protein of haemoglobin which transports oxygen, not round the bloodstream, but into muscle tissue,' recalls Jeffreys. 'To do that work we needed large amounts of myoglobin-rich tissue, which is why we went to the British Antarctic Survey – for whales and seals are world record holders in the amounts of myoglobin they contain. They need lots of the stuff to supply their muscles with oxygen during their long dives.'

Armed with ample supplies of grey seal tissue, the scientists were able to identify the myoglobin gene, and then use that information to find its human corollary. And that is when the surprises began to appear.

The scientists found that tucked away in this obscure human gene was a segment of DNA that was stated over and over again. More to the point, the core of this stretch – which is fifteen base pairs long – was very similar to DNA sequences that other scientists were finding in other chromosomes, and which were also duplicated hundreds of times. Jeffreys realized that our genomes must catch and repeat at certain DNA sequences, just as a person with a stammer will stutter over a word beginning with one letter, say S.

He was excited, for this seemed an ideal, bountiful source of genetic markers for his research on inherited conditions. If a person had a distinctive number of repeats which he or she shared only with relatives, this might be exploited when tracing diseases through families. So on 15 September 1984, using his fifteen-base-pair motif as a probe, which was labelled radioactively, Jeffreys carried out an experiment designed to see if it would pick out the repeated sequences across a genome. By placing them on photographic film, which would be darkened by the radioactive label on the probe, he would be able to see those repeats as black images. And for the DNA samples Jeffreys used specimens provided by a laboratory technician, both her parents, several monkeys, and a grey seal.

'We pulled the film out of the developing tank and saw not only had it worked, but that we had created a DNA fingerprint,' recalls Jeffreys. That single bit of film, faded, fuzzy and still pinned to his noticeboard, told the researchers that the motif could pick out extreme variations, not only among humans but among animals. It could identify an individual absolutely, it appeared – and also establish kinships, for the scientists could see the bands of the technician's DNA appearing among those of her mother and her father.

'It was a very rare occurrence in science,' says Jeffreys. 'It was a blinding flash. In those five golden minutes my research career went whizzing off in a completely new direction. I was channelled away from looking at disease genes, and was thinking about something new – DNA fingerprinting. The last thing that had been on my mind was anything to do with identification, family analyses, forensics, paternity suits and all that. However, I would have been a complete idiot not to spot the applications.'

In short, then, DNA fingerprinting relies on the fact that our genomes each carry a number of sequences that are repeated in ways distinctive to an individual. These repeats can be thought of as a genetic speech impediment. A genome trips up on small, useless sequences (which are sometimes known as minisatellite regions) and, like a stutter, states them over and over again. Crucially, each of us does so in a unique way.

The repeated sequences identified by Professor Jeffreys vary from about sixteen to sixty bases in length, but all have the basic fifteen-base-pair motif embedded within them. Some are duplicated thousands of times, generating stretches of DNA that are more than 60,000 bases long. The trick of genetic fingerprinting is to transform these microscopic diaphanous stretches of biological detritus into hard, black-and-white images that can be seen and analysed.

To do this, DNA must first be removed from a sample of tissue, usually of blood, and purified. The specimen is frozen to −70° celsius, stored and then thawed, a process that breaks down the membranes of the red blood cells (this is known as lysing) but not those of more robust, DNA-containing white cells. Spun down in a centrifuge, the white cells can be condensed into easily removed pellets. An enzyme called proteinase K is added and this bursts open the white cells, dissolving their proteins and eviscerating their DNA coils.

Treatments with phenol and alcohol clear away the white cell's protein debris, leaving behind scraps of precious DNA. At this point one of the crucial ingredients of the genetic fingerprint is added – a special restriction enzyme known as HinfI. We have already seen, in Chapter Four, that restriction enzymes act like biological scissors which cut at special points on the genome. In this case HinfI cleaves at any base pair that is bounded by a GA on one side and a TC on the other. The effect is like placing string in a shredder. The DNA is sliced into tiny pieces – except along one part of the genome, that which is made up of repeated sequences. They do not contain any base pairs bounded by GA and TC and are therefore preserved. What is left are various long stretches of DNA containing repeated sequences unique to the blood donor. It is the length of these stretches, determined by the number of repeats they contain, which characterize an individual.

It should be noted that HinfI is not the only type of restriction enzyme which can be used to make genetic fingerprints. For example, another, HaeIII, will cleave at the point of any GGCC along the genome. Once again, most DNA is shredded except for a category of equally idiosyncratic repeats which has no GGCC sequences, although they are of a similar nature but different classification from those created by using HinfI.

After this, the DNA fragments are separated into bands of different lengths during electrophoresis and are transferred to nylon sheets. Radioactive or luminescent DNA probes, designed to bind to the stuttered sequences, are then added. When they bind, they produce images that form a DNA profile of a human being. Bands shared with parents can be identified and paternity, and maternity, established unequivocally. Similarly, bands shared with those found in blood and semen samples left after crimes have been committed will indicate, with varying degrees of certainty depending on how degraded are the latter specimens, if a person can be included in, or excluded from an investigation.

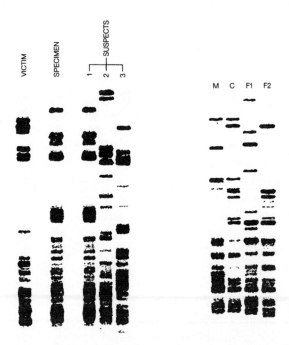

Tell-tale DNA. A genetic fingerprint produces an image that is a unique DNA profile of an individual, providing the forensic scientist with a tool of immense power. The diagram (left) shows how DNA extracted from sperm from a rape victim matches exactly that of one man (suspect 1). Similarly diagram two (right) depicts a mother (M) and a child (C) and shows that the latter shares bands not only with its mother but also with the person (F2) who must therefore be the father.

It is a procedure of formidable precision that has often made headlines since its invention. Yet the test's most sensational forensic use remains its first, a case that had its grim origins on the night of 21 November 1983 when fifteen-year-old Lynda Mann was sexually assaulted, strangled and her body left in a field near her home in the village of Narborough in Leicestershire. The police did not find her killer despite launching an intensive investigation. Then, almost three years later, on the morning of 2 August 1986 the body of a second local schoolgirl, fifteen-year-old Dawn Ashworth, was found in a clump of blackthorn bushes in open land near the village. She too had been raped and strangled.

This time the police seemed luckier. Within five days of the murder they detained a young hospital porter, Richard Buckland, and on 9 August they arrested him for Dawn's murder, after he confessed to the killing. But what about Lynda's death? Had Buckland been responsible for that as well, the police asked? They approached Jeffreys, who had just started using his newly developed brainchild for immigration cases, but not for forensic work. He agreed to help and was sent a specimen of Buckland's blood and samples of semen taken from Lynda and Dawn's bodies.

His brief was simple: confirm that Buckland is guilty of Dawn's murder, and prove that he also killed Lynda. It took a week to set up the test which was expected to show that all the samples had come from the one person, Buckland. It was a nailbiting experience for Jeffreys. This, after all, was the perfect opportunity to show the value of his brainchild and he did not want it botched. In the end, he came into his laboratory late one night so that the experiment could be finished off as soon as possible. 'I just could not wait any longer,' he recalls.

Jeffreys pulled the film from its developing tank and peered at the black bands on it. What he saw gave him a very unpleasant shock. Yes, the semen taken from Lynda and Dawn had come from the same man – which showed that one killer was indeed responsible for both deaths. But no, neither sample matched the DNA in the blood extracted from the young porter.

'I phoned the police and told them. Their language was unrepeatable. They kept asking what the effing hell I was on about – Buckland was guilty. And when, eventually, I told Chief Superintendent David Baker, who was in charge of the inquiry, he simply said: "Yes, very interesting. That means your technique does not work."'

And that thought did occur to Jeffreys. Perhaps some strange mutations in Buckland's DNA had skewed the test. He showed his results to visiting Home Office forensic scientists the next day. 'They basically said: "Don't be silly. It is clear from the DNA. This guy is innocent." They gave me backbone when I needed it.'

So Jeffreys stuck to his guns. Despite Buckland's confession, he was innocent. The police had arrested the wrong man. He carried out a second test
while the Home Office performed a similar, independent analysis. Both confirmed that Buckland could not have carried out the crime. So the police –
still dismayed at losing what seemed a superb conviction – conceded the
case. On 21 November 1986 they presented their findings at the Crown
Court, Leicester – evidence that made legal and forensic history. Buckland
became the first person to be cleared of a murder charge thanks to a DNA
test.

On its own, that was an important accomplishment. However it left the
matter of the real killer's identity unresolved. A £20,000 reward was raised,
but no helpful information was proffered. So detectives decided to adopt a
new and quite revolutionary approach. If genetic fingerprinting had eliminated their chief suspect, perhaps they could use the technique to trace the
real killer, they reasoned. So on New Year's Day 1987, police announced that
all male residents aged between seventeen and thirty-four in the Narborough
neighbourhood would be asked to submit blood samples in order 'to eliminate them from their inquiries'.

By September, more than 4,000 local men had come forward to give
blood, without success – until a chance remark transformed the investigation. During a lunchtime drink with his workmates a Leicester bakery
employee, Ian Kelly, revealed that he had taken the blood test on behalf of
another workmate named Colin Pitchfork. A local resident, Pitchfork had
convinced Kelly – who lived far from Narborough and who therefore had not
been asked to give blood – that the police were trying to frame him. Kelly
was persuaded to pretend to be Colin Pitchfork at one of the 'bloodings' held
in the village hall.

The story worried Kelly's colleagues until one eventually told the police.
Kelly was detained and promptly admitted his deception. Pitchfork was
then arrested. Genetic fingerprints later proved he was Dawn and Lynda's
killer, though by this time Pitchfork had confessed. On 23 January 1988 he
was sentenced to life imprisonment. It was the first and most dramatic
conviction achieved using genetic fingerprinting and was crucial in helping
the technique gain widespread acceptance.

'It was the best possible result,' says Jeffreys. 'If we had failed in flushing
out Pitchfork, the forensic use of genetic fingerprinting would have been
delayed for years. The fact that it worked meant police forces round the
world started screaming for it.

'However, if genetic fingerprinting had blown open the case on its own,
the local police might well have felt belittled by it, and could have become
resentful. As it was, it was a combination of the technology and some bloody

good old-fashioned coppering that cracked the case. We became partners.'

And a very fruitful partnership it has been. Before DNA testing, forensic scientists had only been able to make use of blood types (A, B, AB, and O as well as rhesus negative and other groupings) to differentiate individuals and to try to select people who might be connected with samples left at the scene of a crime. This could provide valuable information on identity but not nearly as precisely as Jeffreys' DNA fingerprinting. Furthermore, it required better preserved specimens than the DNA-based techniques.

Forensic scientists have been able to exploit this new DNA technology in a host of different ways. Apart from exploiting blood, semen and even nasal mucus samples to track down killers and rapists, both hair and saliva have been used to create genetic fingerprints, as is highlighted by the case of Steven Hostettler, an armed robber who was convicted from a genetic fingerprint of his spittle, which he left behind on a security system screen after spitting at it during a bank raid.

Clearly the precision of genetic fingerprinting makes it an extremely powerful forensic device. But the ability to link DNA samples has implications that go far beyond criminal investigations. After all, DNA sequences are inherited equally from both parents. That means that a child's bar code will show if half its DNA has been passed on to it from a particular adult. Apart from settling civil paternity suits, genetic fingerprints have been an enormous boon to immigrants seeking to overturn refusals to be allowed to enter the UK because they could not prove they were the sons or daughters of citizens of this country.

The following example reveals how genetic fingerprints can be used in these cases. A boy, born in the UK, emigrated to join his father. After a period abroad, he returned to Britain to be reunited with his mother, brother and two sisters. However, on arrival, the British immigration authorities claimed a substitution had occurred and he was not the woman's son. The boy was refused entry, and without a means to demonstrate he was her child the woman was powerless, especially as the boy's father was not available for testing. So genetic fingerprints were obtained from the boy, and his alleged mother and siblings. Using bands present in one of the undisputed offspring (but absent in the mother) an accurate DNA profile of the father was reconstructed. The results showed that of thirty-nine paternal DNA bands identified, half were present in the boy's genetic fingerprint. The remaining bands were found to be present in the mother's profile – showing that she was indeed the boy's real mother, and that he had the same father as his brother and sisters. The immigration authorities had to relent, and admit him to Britain.

And stories like these have not been rare. Indeed, one of the most striking features about genetic fingerprinting's first five years of use in Britain, between 1986 and 1991, was its domination by immigration cases. Although forensic applications grabbed the headlines, it was the test's ability to confirm precise, unequivocal relations between living humans that put right the greatest number of injustices. In that period, Cellmark, the company responsible for commercially exploiting genetic fingerprinting in the UK, carried out more than 18,000 tests on immigrants seeking entry to Britain who had been turned back by the authorities. Of those, more than 95 per cent produced results that showed they were blood relatives of UK citizens. These people now have British citizenship, thanks to genetic fingerprinting.

Today, that backlog of cases has been cleared and most genetic fingerprints are carried out to establish paternity and maternity in civil cases, with an additional, substantial minority of tests being done as part of criminal investigations. And it is in this latter area that some controversy has arisen.

To understand the source of this conflict we must first realize that when scientists make a genetic fingerprint they frequently avoid the use of probes that create a full-length bar code of an individual's DNA. Instead, once electrophoresis has separated all the fragments containing repeats into their various lengths, a different type of radioactive or luminescent DNA probe is added. This probe does not bind roughly to all the stuttered sequences (or minisatellite regions) as was described earlier. It adheres to only one sequence. In other words, instead of highlighting all words beginning with S, this probe sticks to only one particular stuttered word, and highlights it and only it. These are known as single-locus probes because they only reveal genetic variation at one location in a person's DNA. Those described earlier in the chapter are called multi-locus probes.

The most obvious difference between a single and multiple locus print is the striking dissimilarity in their appearance. Consisting of dozens of lines of different thickness, the latter looks like a bar code from a supermarket. The former is made up of only two lines, one representing a block of specific repeats inherited from a person's mother, the other from the father. Reducing a genetic fingerprinting to a two-line pattern has several advantages: it is simpler, allows scientists to make comparisons that are less equivocal, and is quicker to operate. Most important of all, it is much more sensitive. A single-locus probe requires fifty times less genetic material than does a multi-locus probe. In more than half the forensic cases that involve genetic analysis, there would simply not be enough DNA to create a profile – a crucial point.

But there are drawbacks. Even if a genetic fingerprint from a bloodstain and one from a suspect produce identical pairs of bars, this does not prove unequivocally that they come from the same person. A given DNA repeat occurs throughout the human population. There is therefore a statistical probability that two people will share a band on a single-locus pair. There is an even more remote possibility that two people will share both bars. And so, to reduce the odds of incorrect connections being made, forensic scientists use three or four different single-locus genetic probes. The end result is a DNA analysis that is quicker to make and clearer in interpretation but which is more sensitive than the multi-locus variety. We should note that it does not uniquely define an individual and so, instead of calling it a genetic fingerprint, we should call these strips of black-and-white imagery genetic profiles.

When dealing with degraded DNA that has been left behind in an old bloodstain or saliva residue, forensic scientists find that it is not always possible to produce several sets of single-locus probes. Two of the probes might not work, and the third might only succeed incompletely, creating a single bar. Nevertheless, this bar might still match one of the lines in a suspect's genetic fingerprint – though the odds that this match might have occurred by chance alone are greatly increased. The forensic evidence becomes less impressive and scientists are often left to make statistical estimates which have been questioned in court.

However, the greatest challenge to the legal use of genetic profiles was made on different grounds and involved the murders, in the early hours of 5 February 1987, of Vilma Ponce and her two-year-old daughter in their Bronx apartment. It was a brutal crime, even by New York standards. Fortunately the police were given an early lead, an anonymous tip-off which led them to Joseph Castro, a local handyman. During examination, bloodstains were found on his watch, which the US laboratory Lifecodes was asked to investigate. It concluded that the stain's genetic profiles matched those from one of the victims' blood. A conviction seemed certain – until a group of biologists, led by Dr Eric Lander of the Whitehead Institute in Massachusetts, stepped in.

When they looked at the genetic profiles made by Lifecodes they found that in one, several bands were shifted out of place, and in others, there were extra bands. Lifecodes had discounted these observations because they thought the former was merely the result of variations in gel composition, and the latter the consequence of slight contamination by foreign DNA, perhaps from a technician.

But Lander pointed to a very different possible cause. It could also be that the sample from the murder scene was simply different in make-up from the sample from the suspect, Castro. It was wrong to claim unreservedly there was a match, he argued.

In the wake of the criticisms from Lander and other molecular biologists, the genetic evidence against Castro was thrown out of court. Ironically, Castro later admitted his crimes, but not before he had helped to tarnish the use of genetic evidence in courts in the United States. Certainly since then a series of dramatic challenges have been made to the legal use of the technique. Most have been refuted, though some persistent doubts remain over one aspect of their employment. These misgivings, which are certainly not shared by all scientists, concentrate on the likelihood of those two single-locus bands occurring with anomalous frequency among certain parts of a population.

To illustrate this point, let us examine what happens when a forensic scientist decides that two bands, one from a crime-scene sample and one from a suspect, are the same. He must then decide what are the chances of that band occurring at random in the general population. If the band is common, the link between suspect and sample is poor. If it is rare, then the connection is strong. Now forensic scientists use four or five different single-locus tests to compare up to ten different bands, so a fairly definite link can be built up. Nevertheless, critics argue that sample sizes of a few hundred people, which are typically used by forensic scientists to establish whether a band is common or rare, are simply not large enough to make sound judgements about a particular band's distribution in a population.

In addition, they argue that some racial groups may be more prone to carrying some bands than scientists realise and therefore will be more likely to produce positive links when none really exists. For example, if a band is very common among Hispanic people (but not other people), then samples left by one such person at a crime could easily, but wrongly, be linked to another person of the same racial background. For their part, forensic scientists argue that they take such considerations into account in their calculations. They have created extensive population databases from all sorts of ethnic groups and have found no evidence for significant profile frequency shifts between them. They also stress the point that was made in the previous chapter – that the overriding amount of genetic variation in our population is due to differences within ethnic groups, not between them. Our genetic homogeneity applies to our DNA fingerprints as much as to the prevalence of genes for blood groups, immune control and the rest.

To settle the issue, the National Research Council in the United States set up a committee on DNA Technology in Forensic Science, which was chaired by Victor McKusick, and which included past critics such as Eric Lander. In April 1992, after going through several redrafts, the committee published its report, which gave a fairly clean bill of health to genetic fingerprinting, subject to a few guidelines. The report concluded that procedures for comparing genetic profiles were 'fundamentally sound' and should be considered reliable – when done properly. And to ensure this last recommendation, it urged that a US government accreditation programme be established to check the quality of forensic laboratories. In addition, to strengthen the statistical basis upon which rested the comparison of genetic fingerprinting, the report urged that researchers take DNA samples from 100 persons in each of fifteen to twenty ethnic groups to create a reliable database.

The committee's deliberations took much of the sting out of the argument about the reliability of genetic evidence, though criticisms are still made, for instance by Richard Lewontin of Harvard University, and by Daniel Hartl of Washington University. They continue to argue that populations contain subgroups within which frequencies of genetic profile markers could vary dramatically.

In Britain where, after all, genetic fingerprinting was invented, there has been very little such debate, mainly because British people are essentially less litigious by nature, and the legal system is less adversarial than it is in the States. None the less, all DNA evidence can be scrutinized by experts for the defence. In addition, there is the problem with the corruption of the term 'genetic fingerprint'. When Jeffreys developed his technology, he used the term 'DNA fingerprint' to refer to his multi-locus probe which produced a bar code that does uniquely identify a person (apart from a pair of identical twins). However, the term has evolved to mean any form of DNA typing, including the use of single-locus probes (which should be called profiles, as we have seen) and others that we shall discuss further on in this chapter. These tests do not have the absolute, rigid accuracy of Jeffreys' original multi-locus probes. 'Lawyers have stood up and said these tests do not have the individual specificity of ordinary fingerprints,' says Jeffreys. 'Then they allege that we are guilty of misleading the public and the judiciary. We are not, and that really annoys me.'

And while we are considering the issue of allegations about the dangers posed by genetic fingerprinting to the innocent suspect, we should also remember the example of Richard Buckland, whom the police originally charged with the murder of Lynda Mann. He might still be imprisoned were it not for genetic fingerprinting. And that is a crucial point. 'About 30 per cent of the suspects on whom we carry out DNA tests are promptly eliminated from inquiries as a result,' says Peter Martin, deputy director of the Metropolitan Police Forensic Laboratory. 'The test works both for and against the prosecution. We should never forget that.' This last view is vehemently supported on the other side of the Atlantic by Dr Edward Blake, a leading scientist and DNA analyst with Forensic Science Associates in San Diego. 'Roughly a third of my work produces results that exonerate men and women who were confidently assumed by authorities to be guilty. Were it not for DNA typing, many of these individuals would have been sent down the river.'

However, there is one definite limitation to the technique – and that is its sensitivity. A relatively large amount of DNA is required before a Jeffreys-style genetic fingerprint can be created. In the case of blood, the usual source of DNA for tests, a few millilitres are required. For paternity and maternity cases that is no problem; such a sample can be drawn without discomfort. But when dealing with crimes in which only spots of blood or decades-old, degraded tissue have been left behind, there is simply not enough DNA to create a genetic fingerprint. Fortunately, this limitation does not affect another technology, that we described in Chapter Four, and which is now being exploited by forensic scientists in earnest. It is the polymerase chain reaction.

As we have seen, a single cell is theoretically sufficient to provide enough genetic material to determine the presence of a DNA sequence. In effect, several dozen are needed for forensic purposes – which still gives PCR two orders of magnitude more resolving power than genetic profiling as originally created by Alec Jeffreys. In other words, it can determine the presence of DNA using samples that are less than a hundredth of the size of those needed for creating a genetic profile. But could that increased resolution provide meaningful information that could match an individual with a sample found near a crime, forensic scientists wanted to know? The answer was provided by a case that matched the Pitchfork murders both for drama and for scientific pyrotechnics.

This story began at 4pm on 7 December 1989 when workmen digging up a back garden at 29 Fitzhamon Embankment, a Victorian terraced house in Cardiff, uncovered an old carpet tied up with electric cable. Inside the rug was a black plastic bag, and within that was a human skeleton. The hands of the victim had been tied at the back with flex.

A murder inquiry was launched, but it was hampered by a very basic problem. Detectives had no idea who the victim was. A variety of experts were called in, and in combination were able to paint a picture that subsequently proved extraordinarily accurate, given the paucity of available information. The person was a female of about sixteen years of age, and had probably lived locally. It was also likely that she had been strangled, and had been buried at least five years before her body was found. Such information was useful but still imprecise. So police asked Richard Neave, an illustrator with Manchester University's medical department, to try to reconstruct a face from the skeleton's skull. Using modelling clay, Neave was able to recreate her features which he built up, muscle by muscle, following the contours of the skull. Photographs were released at a press conference in January 1990.

Two days later a couple of Cardiff social workers telephoned the police to say the face reminded them of Karen Price, a fifteen-year-old girl who had been in care before she went missing in July 1981. 'We have a breakthrough,' announced Detective Chief Superintendent John Williams, head of South Wales CID.

The identification was certainly important, but it was not decisive. Police still had to establish Karen's identity beyond doubt. And that is when they turned to the molecular biologist. Was it possible to extract DNA from her remains, and link this with samples taken from Karen's parents, who were both still alive, detectives asked.

The answer would have been no – had it not been for the development of PCR. Although DNA is found in bone (bone cells have nuclei like the others in our bodies), it is extremely difficult to remove from skeletons that have been in the ground even for only a few years. For a start the DNA degrades noticeably. Of the two types of bases – pyrimidines (cytosine and thymine) and purines (adenine and guanine) – that make up DNA, the pyrimidines, particularly thymine, are susceptible to oxidation. And when this happens, DNA strands crumble and break into pieces. In addition, human DNA becomes contaminated with DNA from bacteria in the soil. Gene amplification was to solve these problems, however.

Part of Karen's skeleton was passed to Dr Erika Hagelberg, one of the

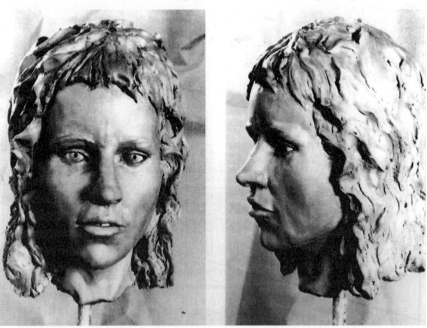

Clay model of the face of Karen Price, reconstructed from the remains of her skull.

Oxford Institute of Molecular Medicine scientists who were investigating the genetics of Polynesian migrations discussed in Chapter Nine. 'When the police called me, I was not even sure in my mind that it was feasible to amplify DNA from bone,' she recalls. 'I had done it a couple of times as part of my research but other scientists, archaeologists and anthropologists, gave me a lot of flak. They said I had probably amplified my own DNA through laboratory contamination. Nevertheless, I thought I would have a go.'

So Hagelberg sandblasted part of one bone to clean it and then began grinding it into a fine powder, eventually producing a sample of 1.5 micrograms of DNA. Unfortunately, 99 per cent of this was found to have come from bacterial contamination. However, with the aid of PCR, Hagelberg was able to amplify the tiny remaining human residue in sufficient quantities to show that it might be possible to attempt an identification.

Now as we have seen, a genetic fingerprint typically exploits stretches of DNA that are many thousand base pairs in length, and which are in turn built up from individual sections that vary in size from sixteen to sixty-four base pairs and which are repeated over and over again. And here lay the essence of the scientists' problem. They found that the bone DNA extracted from Karen's body, which had lain in the ground for eight years, had disintegrated so much that the largest stretches were no more than a couple of hundred base pairs in length. In any case, PCR cannot amplify sequences longer than about 1,000 base pairs and therefore could not generate the necessary DNA lengths required for making a standard genetic fingerprint. So a different form of genetic identification had to be developed.

Fortunately there are many distinctive types of DNA repeats. Those exploited in genetic fingerprinting are large. Others are much smaller; indeed, they are often made up of only a couple of base pairs that are reiterated many times, producing stretches such as CA CA, over and over again.

Hagelberg, in collaboration with Alec Jeffreys, was able to find a number of these sequences. More importantly, she found that many of them varied from person to person in numbers of repeats. And thirdly, and most crucially, the researchers discovered that each of these sequences shared the same flanking region, no matter how many repeats lay in between. This meant that they could assign primers – vital to the process of amplification as we saw in Chapter Four – so that a millionfold copies of a particular repeated sequence could be grown in the laboratory.

For example, when comparing two people it might be found that one individual has thirty repeats of one sequence – say CA – on a stretch inherited from his or her mother, and thirty-four inherited from the father. The second person might have twenty-eight and thirty-two repeats respectively. Now when that section of CA repeats is amplified, the first person would therefore produce a sixty-base-pair product (thirty sets of CAs) and a sixty-eight-base-pair product (thirty-four sets of CAs). The second person would have fifty-six (twenty-eight sets of CAs) and sixty-four (thiry-two sets of CAs). These could then be separated by electrophoresis to produce two pairs of twin bands which would distinguish the two individuals. The effect was almost identical to a single-locus probe. Then the same process can be repeated for a different sequence, say GT, resulting in a different pair of bands. This can be repeated several times for different sequences, to produce pairs of bands, much as it is when creating a genetic fingerprint made out of several single-locus probes. It was this technique that Hagelberg demonstrated on the DNA from the bones of Karen Price.

'I showed the technology was feasible and so we contacted Karen's parents to get blood samples from them,' adds Hagelberg. 'After that Alec [Jeffreys] took over. By amplifying several different sequences from Karen's bone DNA and from her parents' blood, he was eventually able to show that the probability that the skeleton was that of Karen Price was at least 99.9 per cent.'

Armed with this emphatic proof, the police closed in on Karen Price's acquaintances. And during an interview with one, Idris Ali, they got their long-awaited break. Ali broke down and confessed to his involvement in her murder, implicating a second man, Alan Charlton, a builder, occasional nightclub bouncer and convicted rapist. Ali also named a woman who had witnessed Charlton beat Karen to death when she refused to pose naked for photographs. She became the star witness at the trial of the two men. However, it was the strength of the extraordinary scientific evidence arraigned against them that was to convince the jury of their involvement in murdering Karen Price a decade earlier. On 26 February 1991, after a five-week trial at Cardiff Crown Court, Charlton was sentenced to life imprisonment for murder, while Ali was ordered to be detained 'during Her Majesty's pleasure' for helping dispose of her body and concealing the crime.

It was a landmark case, for several reasons. The conviction represented the first use in the world of bone DNA in a forensic case, and the first use of tiny repeat sequences to make an identification. For Europe, it was also the first time PCR had been used in court, the technique having made its debut in America a few months earlier.

But Hagelberg and Jeffreys did not develop PCR typing just to solve the murder of Karen Price. By chance, they had developed the technology to help solve an even more sinister problem, the doubts that surrounded the exhumation of one of World War II's most evil criminals, Josef Mengele. A qualified doctor, good-looking (he was frequently likened to Clark Gable) and in his thirties, Mengele was known as 'the angel of death' at Auschwitz. He would stand at the unloading ramp at the death camp and with a flick of his thumb send prisoners to the gas chambers or his laboratories where they were used as subjects for his 'medical research', experiments that involved surgery performed without anaesthetic, as well as castration and radiation exposure. Even among Nazi war criminals his name evoked a unique odium.

Mengele fled Auschwitz just before it was overrun by Soviet forces in 1945 and disappeared into a network of Nazi sympathizers in South America. For forty years he evaded teams of Israeli Nazi-hunters until he was traced to a grave at the Nossa Senhora do Rosario cemetery at Embu, in southern Brazil. According to locals, Mengele had drowned in a swimming accident in 1979. His exhumed body was examined by pathologists. Both the skeleton and teeth matched records of the fugitive doctor, they said. However, Israeli police chiefs and doctors openly questioned the verdict. The body was not Mengele, they claimed; the man was probably still at large, and had placed another body in the grave marked as his. 'Mengele experimented on humans, especially twins,' said Israeli police chief, Menachem Russek. 'Who could be better to organize such a hoax?'

To settle the issue, Jeffreys was called in. Working with Hagelberg, he was able to identify a tiny amount of DNA from the six-year-old bones that were said to be Mengele's. 'We could only extract about thirty cells' worth of DNA, but that was enough to create a profile using about ten different PCR probes,' says Jeffreys.

Unfortunately Mengele's son, whose blood was needed to provide the final corroborating match, refused to co-operate. And that was in 1990, the year in which the Cardiff police were trying to solve Karen Price's murder. So the probes created for the Mengele case were then turned on her remains, while German prosecutors pressurized Rolf Mengele (or Rolf Jenkel as he now called himself). Under threat of exhuming the entire Mengele family, Jenkel relented and provided a blood sample. It matched the DNA from the Embu cemetery grave with a 99.8 per cent certainty, and the case of 'the angel of death' was closed.

We can therefore see that the forensic power of PCR typing is considerable. It can be used on samples that have only a few dozen cells' worth of DNA in them, and it takes only a couple of days to produce a result. In comparison, the processing of a genetic fingerprint takes about two weeks to complete. It is therefore not surprising that several leading police laboratories, such as the Metropolitan Police's forensic headquarters in London, have begun to use PCR typing, either as an adjunct to, or possibly as a replacement for, genetic profiling. Either way, it is anticipated that the remarkable potency with which PCR exposed the murderer of Karen Price and the identity of the occupant of the grave at the Nossa Senhora do Rosario cemetery, will soon be exploited on a wide scale. Indeed most scientists, including Alec Jeffreys, expect that DNA typing will be completely PCR-driven by the end of the century. In other words, the might of polymerase chain reactions will totally replace genetic fingerprinting of both the single- and multi-locus variety.

And apart from helping to speed up the process of forensic investigations, the spread of PCR typing could have considerable legal implications. As Jeffreys points out, the ability of the polymerase chain reaction to identify key, distinctive regions of DNA from tiny, degraded samples should have special impact on some individuals who have already been convicted of crimes of murder and rape. He believes that while the relatively low-level power of genetic fingerprinting helped prevent many innocent men and women from going to jail, PCR's greater power will actually help free individuals who have already been jailed, wrongly, for serious crimes. Through the clarity of its spectacular magnification, it should be possible to peer through years of injustice to correct wrongs that could be set right by no other means. Indeed, forensic scientists now anticipate a flood of appeals from prisoners who have consistently protested their innocence but for whom, at the time of their conviction, the power of genetic typing was not available. An example of the likely shape of future legal challenges is provided by the case of Gary Dotson, a young school drop-out from a suburb of Chicago, who was charged with raping Catherine Crowell in 1977. Despite his denials, Dotson was convicted and given a twenty-five- to fifty-year jail sentence.

At the time, the case made few headlines. Then in 1985 the 'victim', who had subsequently married and was now known as Catherine Webb, announced on American television that she had lied during Dotson's trial. She never had been raped. She had made up a false description of her attacker and had picked Dotson out at an identity parade to cover up for the fact that she had been sleeping with her boyfriend at the time, a man called David Beirne, whom she feared had impregnated her. But in 1985 Webb became a born-again Christian and claimed she could no longer allow Dotson to languish in jail for a non-existent crime.

The trouble was that very few people, apart from Dotson, believed her. Her recantation was doubted by her original trial judge, by her foster parents, and by the state lawyers who conducted the prosecution of Dotson. In addition, Beirne denied having had sexual intercourse with her around the time of the 'rape'. In the end, however, Governor James Thompson of Illinois bowed to public opinion and freed Dotson on parole, though he refused to grant a pardon. Dotson, still loudly maintaining his innocence, was subsequently jailed again after breaking the strict terms of his parole.

So his lawyers contacted Jeffreys. Could he conclusively demonstrate that semen stains left in Webb's underwear, the principal piece of forensic evidence at the original trial, could not have come from Dotson? Unfortunately genetic fingerprints could not be made of the semen because the samples were too old and degraded, and as Jeffreys had not yet developed the PCR probes which he was to use in the Mengele and Price cases, he was unable to help. So the case was passed to Forensic Science Associates of San Diego, and one of its most senior researchers, Dr Edward Blake.

At this stage, PCR typing was in its infancy, so Dr Blake had to make use of a less powerful, but nevertheless distinctive part of the human genome: the HLA immune system which we met in Chapter Seven. In this case, he used a gene which codes for DQ-alpha, a protein displayed on the surface of white blood cells. 'There is one section of this gene, a segment which is 242 base pairs long, that varies specifically and exploitably from person to person,' says Dr Blake. 'At that time, six different varieties were known, which means, as our genes are arranged in twos, that there were twenty-one different combinations of known pairs of DQ-alpha segments.'

By amplifying this section from the semen stains in Webb's underwear, and from blood from Dotson, Webb and Beirne (the boyfriend), it was possible for Blake to show that the semen was of a type that matched Beirne, but not Dotson. As a result, the US courts decided – a year later – to 'vacate' Dotson's conviction. In other words, they took the view that his original conviction was unsafe, and so Dotson was freed in 1988 – nine years after having been wrongly convicted. It was yet another landmark case in the brief but already highly dramatic history of genetic typing, for Dotson became the first person to have a conviction formally quashed thanks to the skills of the modern molecular biologist. And very shortly we can expect to see many more appeals like Dotson's being upheld, though no doubt as many others will be rejected.

And of course the forensic power of PCR can be used to peer back more than a mere decade into the past. Genetic archaeology is also being revolutionized by the technology's capacity to create millionfold copies of scraps of ancient DNA. Erika Hagelberg is exploiting the technology to study thousand-year-old bones from graves in Polynesia, to help unravel more details of past Pacific migrations. Hagelberg had originally demonstrated the power of PCR by amplifying DNA from bones from graves in the English Civil War cemetery in Abingdon, opening up an entire new field of academic study – the analysis of DNA from ancient bones. Great things are expected of it.

Nor is the technology restricted to analysing human DNA. It can equally amplify genetic material from ancient plants or animals, and while non-human DNA is not directly the subject of this book, study of it can obviously shed much light on our human past. For example, Hagelberg has studied animal leg bones found on the *Mary Rose*, Henry VIII's warship sunk in 1545, which was raised from Portsmouth harbour in 1982. These were shown to have come from a pig, probably the salted variety from the vessel's galley! More ambitiously, scientists at the Hebrew University of Jerusalem are using PCR to try to find human white blood cells inside dead lice attached to combs from ancient Judean graves. They hope to amplify DNA from these cells in order to determine genetic links between ancient and modern Jewish people. And in other experiments, researchers at the University of Manchester Institute of Science and Technology are amplifying DNA from preserved scraps of wheat that are many thousands of years old in an attempt to trace, in detail, the development of Neolithic agriculture.

However, we should not turn away with the idea that PCR is a sort of genetic magic wand. It is a highly effective technique but has its drawbacks. One obvious problem is its startling potency. If a single cell is sloughed from a laboratory worker's skin, if he or she sneezes, or if a pipette is not properly washed, then contamination could ruin a test. Take the example of bone DNA pioneer, Erica Hagelberg. She goes to extraordinary lengths – using only water that is distilled and delivered in sanitized ampoules, for example – to prevent impurities affecting results. Similar care will be needed when forensic laboratories start exploiting PCR on a wide and business-like scale.

In addition, the bands that are produced by PCR probes have less variation than those in single-locus genetic fingerprints. The former have on average only a tenth of the variability of the latter. This affects the statistical power of the technique to differentiate individuals.

There are also drawbacks common to both PCR typing and genetic fingerprints, defects which limit their otherwise extraordinary efficacy. Consider a hypothetical case in which police have a forensic sample – a bloodstain left at a murder scene – in London, and a suspect detained in New York. The former might produce a single-locus probe with bands that are 5.0 and 4.0 units long, while the latter might generate bands that are 5.1 and 4.06 units long. These are not the same lengths, but they might reflect, in a slightly distorted fashion, a latent equivalence. Simple measurement error, due to a number of different factors – gel make-up, or whatever – could be skewing the results. These mistaken appraisals therefore have to be taken into account, limitations that further complicate the use and efficiency of genetic typing. Detaining a suspect in New York and extraditing him to Britain is not an everyday business, after all. Police forces will only go through such a process with a fair expectation of securing a conviction, and that in turn requires solid forensic evidence.

The problem lies with measurement, with actually sizing up the lengths of DNA bands. All sorts of factors affect that process, as we have seen, and can produce tiny variables in results. What is needed is an unambiguous numerical readout, a simple, stark code that would be the digital equivalent of the analogue process that produces genetic fingerprints and PCR typing. And just such a technique, created by Alec Jeffreys, is now being put through trials and may one day become an important component of the genetic arsenal of the modern forensic scientist. It is called Minisatellite Variant Repeat (MVR) analysis.

The key to the development of MVR came with work done on single-locus genetic fingerprints. It was discovered that on a stuttered region, or mini-satellite, called MS32, there are slight variations between the repeated sections. Researchers found there are two versions of a basic, twenty-nine-base-pair long unit that is stated over and over again on one part of chromosome 1. These two variants differ from each other by only one base pair. Crucially, however, the numbers of each variant differ enormously from person to person.

Think of the two variants as black and white beads strung out on a genetic thread. The stuttered region, instead of being made up of dozens of grey beads, is in fact a stretch consisting of a black bead, followed by a white, then black, black, white, white, black, white and so on. This is a simple binary code, which stretches for about fifty repeats – black, white, black, black, and so on – along a chromosome. But of course our chromosomes come in pairs. And the repeated sections on a person's other copy of chromosome 1 will invariably contain a very different string of beads, say white, white, white, black, black, black, white, and so on. This means that when we look at the two stretches of DNA on both chromosomes, we will see that we get parallel pairs of beads, or stuttered variants, and these will come in three combinations: white-white; black-black; and white-black. This is a simple tertiary code which can be turned into a string of numbers from one to three. The number 1 represents a white-white pair; 2 for black-black; and 3 for white-black.

Using PCR to amplify these variants, scientists can then turn the genetic beads threaded along the parallel strands of both our chromosome 1s into a string of numbers: 1, 1, 3, 3, 2, 1, 3, 1, 2, 2, and so on for fifty digits. In this way, a unique digital code for an individual can be created and used to compare forensic samples. If our New York suspect's code should match the blood sample found in London, then this would be likely grounds for extradition. No sizing or measuring would be involved. Only the austerity of a series of numbers would be involved. Of course the technique has its limitations at present – it cannot differentiate the genetic typing of several perpetrators of a multiple rape, for example. Nevertheless it has important potential, because it is unequivocal in its numerical format.

THE MVR™ ANALYSIS CODE

☐ Repeat Unit A   ☐ Repeat Unit T

Father's copy
of DNA repeats

Mother's copy
of DNA repeats

Father's copy	
A	
T	The arrangement of repeat units is different for each copy
T	
A	
T	
A	
T	
A	
A	
A	
T	3rd repeat
A	2nd repeat
T	1st repeat

Magnification of
short section of copy

Mother's copy
A
T
A
A
T
A
T
A
T
A
A
A
A

Father's
Copy

Mother's
Copy

The MVR™ test records
the repeat unit order in
both parent's sets
simultaneously. This
combination results in the
code for the individual.

Father's Copy	Mother's Copy		
A	A	=	1
T	T	=	2
T	A	=	3
A	A	=	1
T	T	=	2
A	A	=	1
T	T	=	2
A	A	=	1
A	T	=	3
A	A	=	1
T	A	=	3
A	A	=	1
T	A	=	3

**THIS IS THE
MVR™ CODE**

A	+	A	=	1
T	+	T	=	2
A	+	T	=	3

There are two versions of a
repeating unit region (minisatellite),
one inherited from the mother, one
inherited from the father. Each
minisatellite is made up of two types
of repeat unit and can be hundreds
of units long.

This person's code is: 3 1 3 1 3 1 2 1 2 1 3 2 1

'The critical advantages of MVR lie with the fact that it is driven by the gene amplification, by the technology of the polymerase chain reaction, which therefore means it can be used effectively in the analysis of the tiniest of samples,' says Dr Paul Debenham, research director of Cellmark UK, which has been developing the technology. 'It also manages to decode tremendously informative repeat variations in a single test.

'In addition, the technique involves no standardization of procedures between two laboratories. As long as two centres can differentiate between the black and the white repeats, that is all you need. You can compare your results numerically – by making a phone call or sending your data down a computer link.'

And it is this last prospect that does cause concern. Some people fear that we might one day create a national, or even an international, database of MVR numbers, or some numerical code created by DNA typing. They see the collation of such information as the thin edge of a genetic wedge, a grim precursor of a type of society outlined in Huxley's *Brave New World*.

Is this a realistic prospect? The answer is yes, it is. The technology is ripening, and the end results could undoubtedly be extremely fruitful. Armed with a computer network stuffed with MVR numbers for every citizen of a nation, it would be extraordinarily easy to run cross-checks and correlations for each new number generated from a bloodstain found at the location of a murder or assault. On the other hand, the implications for civil liberties are profound and few scientists believe such a scheme will ever be contemplated by any modern democracy that has the sophistication to implement the technology.

So where does the future of DNA typing lie? We have seen how it can unequivocally match two tissue samples, one from a murder scene, the other from a suspect, for example, with spectacularly effective outcomes. But what if no suspect is immediately available? Could forensic information from DNA actually help build up a picture of an assailant? Will we one day be able to create DNA photo-fits of suspects?

The answer is yes, but only up to a point. We can already determine the sex of a person who has left tissue such as blood at the scene of a crime. Within a few years, however, it should also be possible to make pronouncements about some fairly obvious physical characteristics of that person – his or her hair and eye colour, for example. It should even be feasible to determine ethnic origins from a blood or semen stain, though only with a probabilistic, and not an absolute certainty. In other words we will only be able to say there is a fifty times higher than average chance that the person is a Bengali, or a Welshman, or whatever.

'The trouble is that we know nothing about the genes that control appearance,' says Jeffreys. 'All sorts of different factors are involved in something as simple as the length of our noses.' However, we do know from studying identical twins that a large part of the acquisition of facial characteristics is determined by our genes. A careful analysis of these features combined with family studies and gene mapping techniques could one day help locate those highly variable genes that are responsible for the human face. The day of the DNA photo-fit would then be upon us. And when it is, we may find our faces to be 'as a book where men may read strange matters', to quote Macbeth.

The story of DNA fingerprinting, 'accidentally' conceived as an adjunct of basic genetic research but which has rapidly become one of forensic science's most valuable techniques, gives some idea of the general impact that future molecular biology will have on society. Nor does the story stop here. In the case of DNA typing, we might be able, sometime in the next millennium, to paint a picture of a suspect just from a microscopic drop of his or her blood, despite all the difficulties involved. And in many other facets of genetics, similar startling developments will be made. Let us now look at the most spectacular of all these advancements – the alteration of the molecular make-up of a living human being: gene therapy.

11

The Splice of Life

Since 1991, Cynthia Cutshall and Ashanthi Desilva have spent their days going to school, playing with their friends, shopping with their parents and generally passing their time like typical American youngsters. It sounds unremarkable. But there was a time when such an existence seemed an improbable, distant dream; a time when life for Ashanthi and Cynthia was filled with bouts of interminable infections, followed by spells of quarantine at home. So devastated were their immune systems, and so powerless were they to fight infections, that neither girl was given much chance of living beyond adolescence.

Then, within a few months, their lives were transformed, thanks to a remarkable piece of pioneering medicine which exploited the techniques of genetic engineering in order to bolster their stricken immune defences. Ashanthi and Cynthia became recipients of the world's first trials of gene therapy. An entirely new type of medicine had been created, one that at worst should alleviate several rare, inherited, life-threatening conditions, and at best may one day be used to treat congenital heart conditions, Aids and cancers, including those of the brain, skin and breast.

In a sense, gene therapy represents the pinnacle of molecular biologists' achievements. In a few decades they have learned to pinpoint defective genes, sequence them and their normal counterparts, and create new drugs using the principles of cloning. Now, with the advent of gene therapy, they have actually begun to alter the genotype of a human cell. This technology is based on several scientific milestones that we have already discussed – including the development of cloning and the unravelling of the behaviour of viruses. However, there were other, more personal involvements, and one of the most striking was the commitment of Dr Kenneth Culver.

In the late 1980s Dr Culver had tried to save the life of a young girl called Chelsea who had been diagnosed as a sufferer of the same disease that struck Ashanthi and Cynthia, an immune ailment called adenosine deaminase deficiency, or ADA deficiency. 'The white blood cells that patrolled Chelsea's body were ill-equipped for their job; a large contingent of that police force, the T-cells, manufactured a poison that caused the cells to self-destruct,' as he states in a review of gene therapy in the journal, *The Sciences*, published in 1993.

As a result, Chelsea could not fight even simple infections that healthy bodies mop up without effort. While still only a few months old, she had to be given a bone marrow transplant from her father in a bid to rejuvenate her flagging output of lymphocytes. She also received chemotherapy and was confined to a sterilized hospital room to be tended only by people in scrub suits, masks, hats and gloves. It was to no avail. Chelsea continued to sicken. So a second bone marrow transplant was arranged, this time from her mother.

'For almost two weeks we watched anxiously for signs that the transplant had taken,' says Dr Culver. 'But the outlook was grim. The white cell count was flat, and I realized that the operation had failed. Chelsea was going to die. As I waited with her parents for the last heartbeat, we felt utter anguish and despair.' Chelsea was eleven months old when she died. Shortly afterwards, however, Dr Culver joined the National Institutes of Health (NIH) in Bethesda, where he was asked to join a revolutionary new programme, led by Dr French Anderson and Dr Michael Blaese, that aimed to cure ADA deficiency using the newly developed techniques of molecular biology. It was the perfect chance to channel that despair.

ADA deficiency is one of a category of inherited disorders in which a person's immune system is found to be incapable of challenging infections. Its most famous victim was David, who was portrayed by John Travolta in the 1976 television film *The Boy in the Plastic Bubble*. David eventually died when he was given bone marrow from his sister, a transplant that unwittingly contained cells infected with the Epstein–Barr virus which rapidly ravaged his body with cancer. Within a few months David died, his liver, lungs, intestine and brain riddled with tumours.

Fortunately, severe combined immune deficiencies (Scids) are very rare. In the United States, for example, there are only about fifty victims born each year, individuals who are prey to a constant stream of infections, often of the ear and throat. Few live past the age of four. And in every case the basic defect is traced to lymphocytes. In particular, from victims of ADA deficiency – one of the main types of 'Scid' disease – it was found that the cause of the condition lay on chromosome 20 on which rests the gene responsible for manufacturing adenosine deaminase. This gene occasionally undergoes mutations; sometimes a tiny sliver of DNA is excised, now and then a fairly hefty section is missed out. For carriers with a normal gene on their other chromosome this is no problem. But occasionally a child inherits a mutated gene from both parents. In nearly every case, the result is disruption to the body's output of adenosine deaminase, an event that, in turn, unleashes a dreadful chemical cascade. Adenosine deaminase normally breaks down a chemical called deoxyadenosine. In ADA-deficient patients, however, deoxyadenosine builds up alarmingly. Most cells in the body can deal with this accretion except, for unknown reasons, the T-lymphocyte cells, which are poisoned and killed off. Without the presence of the body's molecular police, its villains – germs and microbes – promptly run riot.

Once this chemical chain reaction had been uncovered, it suggested an obvious cure to scientists. Simply inject the missing ADA into patients. Unfortunately, because the enzyme survives on its own in the body for only a few minutes, this did not work. Coating the ADA with polyethylene glycol (PEG) increased its survival time; however, treatments typically cost between £50,000 to £100,000 a year and only some patients respond significantly. Similarly, tissue-matched transplants of bone marrow – in which our T-cells are manufactured – have also helped in some cases. Nevertheless, there remained a sizeable fraction of young patients for whom medicine could do little.

And that is where the first gene therapists stepped in. They wanted to find out if they could correct inherited ailments, a task that only a few decades ago would have seemed quite impossible, but with the development of modern molecular biology now seemed feasible. And they chose ADA deficiency for their first trials. This decision was based on several criteria. For a start, the normal ADA gene had been isolated and cloned. In addition, the body needs very little of the enzyme to put right its immune problems. Most importantly of all, however, the ADA-deficient children who were finalized as candidates for gene therapy had failed to respond to all other forms of treatment. Their prognosis was grim. In the end, two children, Ashanthi and Cynthia, were selected because their T-cells, upon which the therapy relied, grew well in cultures. So the trials went ahead.

But first of all doctors had to find a way to deliver the ADA gene into the two girls' T-cells. They chose retroviruses because these infect cells and insert their own genetic information in the process. (We encountered retroviruses in Chapter Six. They are made of RNA, and after entering a cell through its membrane they make a special DNA copy of themselves. This copy is then inserted into the host DNA. Examples of retroviruses include the Rous sarcoma virus, and HIV, the cause of Aids.) More importantly, retroviruses multiply by attacking dividing cells. There is an associated danger, however. Sometimes retroviruses trigger cancer in this way. To get over these difficulties, researchers found that by stripping out three genes from a retrovirus, its carcinogenic properties seemed to be neutralized. And in place of these researchers inserted the ADA gene.

The aim of this experiment was simple. By placing the ADA gene inside a retrovirus, scientists had devised a way to transport the missing gene into a vector – as the neutralized virus is called – which in turn would carry it into the genomes of the girls' crippled T-cells. Of course there is no way to guide this insertion. The new gene is lodged by chance on any chromosome. However, once there, it manufactures the adenosine deaminase and so prevents the T-cells from committing cellular suicide. At least that was the theory.

Tests on mice and monkeys were encouraging. But would it work on humans? Such was the desperate plight of Ashanthi and Cynthia, that the ethical committees of the NIH and the federal Food and Drug Administration gave approval. So did the girls' parents. Ashanthi – then aged four and the worst affected – was treated first. Her therapy began on 14 September 1990 at the pediatric intensive care unit of the Clinic Center of the National Institutes of Cancer, when her white blood cells were removed by apheresis. 'In that process, blood flows out through an intravenous tube and the white blood cells are extracted,' says Dr Culver. 'The red blood cells are then immediately infused back into the body to prevent anaemia. The white cells are put in culture, and the disabled mouse retrovirus, bearing the inserted normal ADA gene, is added to the cells once they begin to grow.'

The T-cells, now healthily complete with effective ADA genes, were returned to Ashanthi. 'She received a billion cells,' says Dr French Anderson. 'The infusion took twenty-eight minutes. The entire undertaking was clinically uneventful. And so began human gene therapy.'

Four months later it was the turn of Cynthia, who was aged nine. And in both cases, it was found that the girls' altered T-cells survived. 'The number of T-cells in their blood jumped from low levels to normal,' says Dr Culver. 'As a result, the ADA activity we measured in the T-cells increased to 10 per cent in Cynthia's case, and to 25 per cent in Ashanthi's. That may not sound like much but people with 10 per cent of average ADA activity can have normal immune function.'

Antibody levels – another measure of the immune system's strength – also improved, while skin testing also suggested immune function improvements. Before their gene therapy, Ashanthi and Cynthia produced little skin reaction when antigens such as tetanus or diphtheria toxins were injected

Human Gene Replacement Therapy

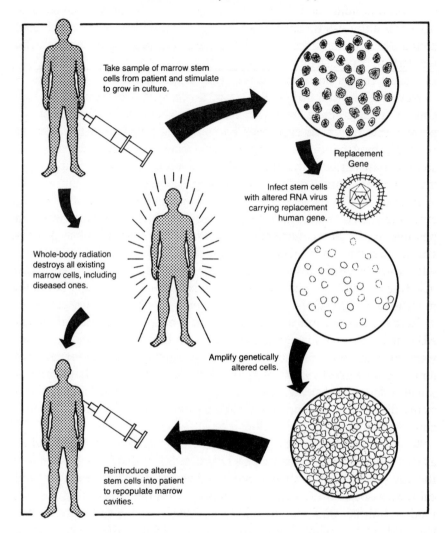

Take sample of marrow stem cells from patient and stimulate to grow in culture.

Replacement Gene

Infect stem cells with altered RNA virus carrying replacement human gene.

Whole-body radiation destroys all existing marrow cells, including diseased ones.

Amplify genetically altered cells.

Reintroduce altered stem cells into patient to repopulate marrow cavities.

under their skins. In healthy individuals, a red bump forms as T-cells invest the area to fight the poison. After therapy, the girls began to react to many antigens, though not all of them.

But perhaps most remarkable of all were the anatomical changes. 'Both Ashanthi and Cynthia were born without tonsils, which are storage sites for lymphocytes,' recalls Dr Culver. 'Within a few months of gene therapy, tonsils of normal size could be discerned in their throats.' Ashanthi and Cynthia now 'lead essentially normal lives', as Dr Culver says in his review. 'Ashanthi's infections began to drop significantly after three infusions of T-cells over a six-month period. Soon she was getting infections no more often than her usually healthy sister. Ashanthi is no longer homebound; she has just completed a year of kindergarten at a public school, and she leaves the house regularly to dance, swim and ice skate. As for Cynthia, her chronic sinus disease and frequent headaches have all but disappeared.'

Nevertheless, the girls still have to go through regular gene therapy courses, because their T-cells are naturally replaced every few months, a process that occurs in everyone's body. And of course, the replacements produced by Ashanthi and Cynthia's bone marrow cells do not make ADA, so they must periodically undergo the same process of T-cell 'rejuvenation' involving harvesting, viral infection and re-injection.

In other words, the gene therapy given to Ashanthi and Cynthia is not a cure, although two years later it appears to be an effective treatment. So is it possible to put their condition right permanently? The answer is probably yes, and trials began in Milan in 1992, and London in 1993, as well as at the NIH, with the aim of achieving this. The goal is straightforward: instead of inserting the ADA gene into T-cells which have a fixed lifetime of only a few months in the blood, researchers now want to correct their stem cells.

Stem cells are the precursors of all blood cells, red and white. And if their ADA deficiency could be rectified just as Ashanthi and Cynthia's T-cells were put right, then a lasting way to restore patients' maimed immune systems could be developed. This is the methodology that was adopted by the Milan, London and Bethesda teams. In each case, the hope is that altered stem cells will find a home back in the patients' bone marrow and there form a more permanent residence for effective T-cell manufacture.

It sounds encouraging but there may be some dangers to the procedure, as Dr French Anderson makes clear in a critique of the general use of gene therapy. 'There are two categories of risk,' he says. 'First there are the clinical risks associated with any cell infusion: infection, chills, fever, etc. In a modern hospital with good blood banking procedures and competent medical and nursing care, these risks are low. Second there are the risks associated with the gene transfer procedure itself. The possibility of inadvertently producing an infectious virus that could harm the patient and the public has been reduced to near zero. What remains as a finite risk, however, is the potential for inducing a cancer. The vector carrying the new gene is integrated randomly into the genome. This means that there is a theoretical possibility that vector insertion may activate an oncogene or inactivate a tumour suppressor gene.'

But as Dr Anderson also points out, and as we saw in Chapter Six, cancer is a multistep process, and an activated oncogene would be but one step along the oncogenic pathway. 'Extensive safety studies have been carried out over the past five years in animals – mice and primates – and not a single vector-containing tumour has yet been found. Nonetheless, the finite possibility of a cancer caused in part by the gene transfer procedure exists. This risk must be balanced by the potential benefits.'

The important point is that for the foreseeable future, gene therapy is likely to be restricted only for use on extremely ill patients. Certainly, for Ashanthi and Cynthia and their parents, the transformation they have gone through would have seemed a forlorn fantasy only a couple of years earlier. With the creation of effective gene therapy treatments, and with the prospect of permanent ones being developed, that position has been changed for ever. But can the lessons gained from adenosine deaminase therapy be generalized to other conditions, ones that are far more likely to strike the average person? In other words, could we one day exploit gene therapy to treat cystic fibrosis or cancer or heart disease? The answer may well be yes, though the extent and permanence of gene therapy's efficacy, in each case, will take years to establish.

Consider the example of cancer. By 1993, researchers at several hospitals and biotechnology companies had begun investigating ways to carry genes into tumour cells to kill them directly or render them vulnerable to normally innocuous chemicals. One of the most ingenious of these attempts was developed by Dr Culver at the NIH, while working with Genetic Therapy Inc. of Gaithersburg, Maryland. The particular cancerous targets selected for this approach are brain tumours.

The system developed by Genetic Therapy depends on the retrovirus system that was used to create Ashanthi and Cynthia's ADA therapy. However, instead of carrying the adenosine deaminase gene, the virus carries a gene from the herpes virus. This segment of DNA codes for an enzyme called thymidine kinase which renders the tumour cell vulnerable to a drug called Ganciclovir, which is used to treat herpes virus infections. Quite simply, a cell which makes thymidine kinase is killed.

The strategy then is to make tumour cells susceptible to a standard treatment – for herpes; though this tactic does beg the question of why the retrovirus should only infect, and insert the gene into, brain tumour cells rather than all brain cells. 'The very nature of retroviruses and brain cells makes that straightforward,' explains Dr Marc Schneebaum, vice-president of Genetic Therapy. 'Retroviruses only infect cells that are in the process of dividing and growing. However, after early childhood our brain cells stop doing this. Our total complement is fixed from then on. That means that the only brain cells that will take up the genetically altered retrovirus are dividing tumour cells. That is how we intend to get a gene which makes a cell vulnerable to Ganciclovir into a tumour and only a tumour cell.' By 1993, treatments using thymidine kinase genes and Ganciclovir had begun on several patients. Definitive trials of the technique are likely to take several years, however.

Then there is heart disease, the number one 'killer' of people in the West. As we will see in the next chapter, there are quite clear signs that some components of the complex biological processes that leave us vulnerable to such ailments are inherited. And one of the most important is familial hypercholesterolaemia, a simple, inherited trait in which victims suffer a build-up of very high blood levels of cholesterol caused by a defect in cell receptors which take up lipids called low density lipoproteins (LDL). Familial hypercholesterolaemia comes in two forms. In the first, a patient lacks a single functioning gene that codes for the LDL receptor protein. It is the job of this receptor to mop up excess of cholesterol in the blood. Without such a gene operating on one chromosome, a person's power to eradicate cholesterol is halved. He or she has to work on genetic 'half-power', with only one normal gene coding for a protein that eliminates excess cholesterol. Such individuals are left prone to heart attacks and coronary disease by the time they are in their late forties.

That is bad enough. Occasionally, though, two of these victims will meet and marry, and produce a baby that will inherit a malfunctioning gene from both parents. By the time such a child is five or six years old, he or she will produce the symptoms of heart illness that are normally seen only in an adult who has smoked, drunk and eaten too much for decades. To save these children, doctors either have to carry out a transplant of a liver, where the LDL receptor gene is made, or complete blood transfusions every two or three weeks. Few patients live beyond the age of thirty.

One such victim, a twenty-seven-year-old French-Canadian girl, was the subject of the first gene therapy attempt to correct this defect. In June 1992, researchers at Michigan University at Ann Arbor used genetically altered retroviruses to carry the missing LDL receptor gene into her liver. Within six months they reported 'extremely encouraging results'. Just how effective the long-term prognosis will be for other serious familial hypercholesterolaemia victims remains to be seen. Once again, though, a start has been made.

Then there are the commoner inherited diseases, of which the most attractive target is cystic fibrosis. Despite the discovery of the basic protein pathway that produces its symptoms, the development of medications for cystic fibrosis remains elusive. So several teams of researchers are trying to find ways to insert the correct version of the cystic fibrosis transmembrane regulator, or CFTR, into cells – mainly from the lungs – of victims. Some groups are using the normal retrovirus approach of the ADA, brain tumour and familial hypercholesterolaemia researchers. Others, such as Professor Williamson of St Mary's Hospital, London, are adopting more radical measures. 'We want to avoid using viruses, and the associated problems of their random insertion of genes into host cell genomes,' he says. 'Instead we want to make a package purely out of human genes – an artificial chromosome. However, the problem is to avoid the defences of the cell, which would usually attack and destroy the gene. So we are going to wrap it up with human proteins that will both protect it and help it gain entry to just the right cells in the lungs of a cystic fibrosis patient. If it survives, it could then start to make the correct version of CFTR.'

In all, there are several dozen gene therapy programmes aimed at correcting the worst excesses of a variety of serious ailments that are the outcome of those losing throws in the dice of life – inherited ailments, heart disease and cancer. Many of these trials could even lead to the development of effective treatments one day. But the problems of inserting a gene into its proper place in a patient's genome, and ensuring that it codes effectively for sufficient levels of a missing protein, are likely to remain significant headaches for gene therapists for many years. The potential is immense, the track record limited.

Not that researchers are limiting their concerns to altering the genomes only of human beings. They are also adding and removing genes from animals, an approach that will also have profound consequences in the creation of new medicines for human diseases. However, in these cases we cannot use the term 'gene therapy' to describe the procedure, for the aim is not to correct a life-threatening mutation, but to introduce one, quite deliberately, into the biological make-up of an animal, usually a mouse.

By inserting genes which, in a mutated form, cause inherited human ailments (again cystic fibrosis is a favourite target) or which are critical initiating steps in triggering tumours, researchers can create animals on which drugs can be tested to treat these diseases. Such creatures are called transgenic animals.

'Before transgenic animals were developed, there were two standard ways of testing a drug or other treatment,' says Professor Christopher Marshall of the Institute of Cancer Research in London. 'We could work on tissue cultures in the laboratory or we could inject an animal with a tumour and then treat it. The problem with the first method was that it dealt with a very artificial situation. A tumour is not dangerous until it is vascularized – which means it sets up a nutrient system that sustains its growth. And you could only create that in living animals which are the only means we have for fully testing anti-cancer drugs.'

But to create an animal with cancer, researchers – until the development of transgenic mice – had to inject them with tumours. Then they would try to cure them. Using transgenic animals instead is a much more precise experimental procedure, and provides much more information which in turn can be used to improve cancer treatments.

For example, we saw in Chapter Six that cancer is caused by several independent genetic mutations in a cell. These can take years to occur together. By introducing one mutant cancer gene, however, the other stages can be induced more speedily, and the other initiating causes of cancer uncovered. 'By using transgenic animals we can find the other genetic slip-ups involved in cancer,' adds Professor Marshall. 'This is an important way forward for cancer research.'

Examples of transgenic animals include mice that have been implanted with mutant genes implicated as triggers of pancreatic and colon cancers, and treatments that are being studied to correct these conditions include boosting animals' white blood cells with drugs like interleukin-2 to stimulate immune attacks on tumours. In addition, a strain of mice has been created that carries the gene that causes cystic fibrosis.

Of course, the creation of animals that are 'doomed to die' of cancer or cystic fibrosis does upset some people, as does the prospect of tampering with the genetic make-up of a human being – which is of course, the raison d'être of gene therapy. Both the technologies discussed in this chapter there-fore raise ethical problems and have caused controversy, issues that are discussed in the last chapter of this book. The main criticism is that in embarking on techniques like gene therapy, scientists have launched them-selves down a slippery incline of moral descent, in which we may end up creating and breeding 'designer humans'. However, those who question whether we should go near that slope at all need only look at the normality of the lives of Ashanthi and Cynthia today for an answer. It was clearly worth beginning the effort. Where we stop is a different question.

12

Mapping Our Genes:
A Personal History
by Walter Bodmer

Max Perutz (the Nobel Prize-winner whom we encountered in Chapter Three as one of the luminaries of the Cavendish Laboratory) once suggested that a future examination paper might one day include the question, 'Prescribe a therapy at the molecular level for Hamlet'! Now we may not be ready to think of providing such a treatment for the poor prince's unhappy mental state at the moment but, as we have shown in this book, science is at least finding answers about the basis for some inherited misery. So far, this mostly involves single-gene ailments such as cystic fibrosis and muscular dystrophy. But we are also beginning to obtain fundamental understanding of more complex disorders, particularly cancers. We can now even think how we might use that knowledge to detect, prevent or cure these conditions.

And that dream will come about through the complete interpretation of the Book of Man, the sequencing and cataloguing of all genes that lie along the 24 chromosomes (22 plus the X and the Y) into which the human genome is divided. This information will reveal the functions of genes and their contributions to human diseases, and will lead to new approaches to their prevention and cure. It will also generate a new understanding of the nature of the human species, its exquisite variation, its evolutionary origin and its recent history. The elucidation of the Book of Man is the aim of the world-wide Human Genome Project, and in this chapter I shall trace my own personal path towards this goal.

My father was a Jewish doctor, and he desperately wanted at least one of his three sons to follow in his footsteps and study medicine. Unfortunately for him, my elder brothers opted for physics and music respectively, while I drifted into mathematics. However, I was rescued from an uncertain future as a mathematician by an interest in statistics when I was a student at Cambridge, and that led me to the great statistician and geneticist, Sir Ronald Fisher, then professor of genetics at Cambridge. He was world renowned, not only for pioneering a quantitative theory of evolution based on Mendel's principles, but also for being the founding father of modern statistics. Thanks to him, we now conduct rigorously designed experiments upon which major advances in agricultural productivity have been made and upon which the testing of new drugs is based today. I went to do my PhD research – on the mathematical study of the behaviour of genes – with Fisher, as one of his last students.

Fisher was one of the great intellects of the twentieth century and he had a profound influence on my career. It was from him that I first learnt about the discovery of DNA's structure. I remember him, at a lecture, taking out of his pocket a crumpled copy of a paper that Watson and Crick had given describing their work. Then he gave a concise description of DNA's structure before plunging into apparently unrelated mathematical formulae. And it was from Fisher that I learned the importance of immersing myself in experimental work and analysing my own data. One of my first tasks was to explore the population genetics of the 'incompatibility system' of the common wild primrose. There are two common forms of the primrose, easily distinguished once you look closely, called pins and thrums. They differ by the position of the anthers which contain the pollen and the stigma which receives it, and only matings between pins and thrums are effective. In some parts of England, however, there is an abnormal, or should one say unusual, form called homostyle, which places the stigma and the anthers in the same position and which is self-fertile. My task was to explain the pattern of distribution of these differences in primula populations, just as blood group distributions differ in human populations.

It was my first introduction to a gene cluster, which determines the differences between pins, thrums and homostyles, rather like the rhesus or HLA blood group systems. The work also introduced me to the subtleties of the behaviour of genes that lie close together on the same chromosome. Indeed, it was the common primrose that gave Fisher his fundamental insights into the genetics of the rhesus blood group system. In turn, I drew inspiration from Fisher when working with my wife, Julia, on the immune system, research that led to the discovery of the HLA system, and to an understanding of the association between HLA tissue types and diseases, which in turn provided the genetic marker clues through which the cystic fibrosis gene was eventually found.

Now we have already seen throughout this book that new insights often have the most unusual origins, and may come from observations that seem at first to bear no relationship to each other. Who would have thought that studying differences in the wild primrose populations could have anything to do with cloning the gene for cystic fibrosis, for example? This shows why it is so important to nurture a scientist's innate curiosity without demanding an immediate application for his or her work. Of course that does not mean that goal-oriented research is not worthwhile. In fact, the path between basic and applied research is tortuous, and certainly the ultimate goal should never be far out of sight.

My problem as a research student in genetics at Cambridge was that I had no biological background. That meant that I not only had to learn about DNA's structure but also about modern molecular biology and its involvement with moulds, bacteria and viruses. I went to Francis Crick to seek advice on how to become a molecular biologist and, given my mathematical background, he suggested I try to find out how DNA's four-letter language codes for the twenty-letter language of proteins. But I could see no basis for a deductive solution to this problem. (I was right, since the answer came from Francis Crick and Sydney Brenner's elegant genetic experiments, not from deductive reasoning, as we saw in Chapter Three.) Instead I went to train in experimental molecular biology with Joshua Lederberg of Stanford University Medical School. Lederberg had won the Nobel Prize for showing that bacteria can reproduce sexually, and in doing so laid the foundations for the science of bacterial genetics and ultimately genetic engineering.

So in the early summer of 1961 I took a month-long 'phage course' at Cold Spring Harbor on my way to Stanford. Cold Spring Harbor was a Mecca for molecular biologists, and the course was specially designed to teach molecular biology to neophytes from other disciplines. It was there that I met many future colleagues and friends. I particularly remember sitting at breakfast next to one young German who told me he had just completed an experiment in which he had added the RNA sequence UUU (U stands for uracil, the RNA equivalent of DNA's thymine) to his magic extract for synthesizing proteins and out had come the totally unexpected and unusual synthetic protein, Phe Phe Phe, where Phe stands for phenylalanine, one of the twenty amino acid building blocks of proteins. The young German, Heinrich Matthei, had discovered the first element of the genetic code – that UUU codes for phenylalanine.

At this time, Stanford University Medical School was collecting Nobel Prize winners. In addition to Lederberg, there was Arthur Kornberg, who had discovered the first enzyme that assembles DNA. Lederberg and Kornberg decided to try to synthesize DNA which could be shown to have biological powers. I was given the job of acting as a go-between for the two laboratories. My task was to test the DNA made in Kornberg's laboratory for its biological activity. I learnt a lot about DNA, how to purify it and how to make one of the only two enzymes that were then known to cut its wispy strands. But the life-synthesizing experiments were frustratingly unsuccessful, for reasons that were not then clear. The answer was simply that not enough was yet known about the DNA duplicating enzymes. Nowadays the experiment we failed to make work is a common, everyday part of the technology of genetic engineering, and underlies much of the science that we have been discussing in this book.

Lederberg and Kornberg are two very different sorts of scientists, each outstanding in his different way. Lederberg is the broad thinker who works on a wide canvas. He is an inspired speculator, though sometimes perhaps with more flair than care. Kornberg on the other hand is a careful experimenter who persists in pursuing a defined goal with forethought and precision. Extraordinary imagination certainly was needed even to think that it might be possible to isolate an enzyme that could copy DNA in a test-tube. But it was dogged persistence that produced a result. In this way, Kornberg is in the mould of Fred Sanger or of careful experimenters like Rosalind Franklin and Maurice Wilkins who provided the crucial X-crystallography pictures of DNA from which Watson and Crick drew their inspiration. Watson and Crick, on the other hand, fit more into the Lederbergian mould of brilliant speculators, though their results were substantiated by a great deal of careful and precise model building.

Human genetics at that time seemed crude compared with its elegant bacterial counterpart. The problem was that you could not make crosses between people and even if you could, you would have to wait for years for the results! Even then, you could hardly have matched the numbers of observations that Mendel had made on his pea plants. The solution to this dilemma was provided by Guido Pontecorvo, who had come to Britain from Italy in the late 1930s to escape fascist persecution of the Jews. He developed an ingenious system for genetic analysis of the mould aspergillus, and he believed this could be applied to human cells grown in tissue culture. Now I had spent six weeks in Pontecorvo's laboratory in Glasgow to learn about his approach to genetics before going to Stanford, and had learned about his idea of doing human genetics with cell cultures. The concept continued to lurk at the back of my mind when I moved to Stanford.

As we saw in Chapter Seven, establishing human tissue types has been based on work with human white blood cells. Then, in the early 1960s, it was discovered how to produce hybrids between different cells in tissue culture. Here at last were all the ingredients needed to do somatic cell genetics. Cells could be grown conveniently in the laboratory, and white blood cells could be obtained from any individual by taking a blood sample. Hybrids between these two different types of cells could now be made and tissue types and enzyme differences could be studied in the cells cultured in the laboratory.

At this time, Mary Weiss and Howard Green published their classic description of crosses between human and mouse cells, laying the foundations for doing human genetics in the laboratory. We pursued this line by crossing human lymphocytes with a suitable strain of mouse cells. This experiment worked because the human–mouse hybrid cells tended to throw out many, but not all the human chromosomes, but keep the mouse ones. The results were extremely useful, for it meant that any given hybrid cell produces a colony of cells that carries a subset of human chromosomes, sometimes only one. The particular subset that is present will vary from one hybrid to the next. Once the individual human chromosomes present in a hybrid could be identified, we could then ask: what human chromosome is always present when a particular gene is present in a hybrid? In this way a gene could be assigned to its chromosome, which is how my laboratory first allocated genes for human enzymes to the human X chromosome, and showed that the genes for a pair of enzymes not previously mapped were present on the same chromosome. Of course, if there is only one identified

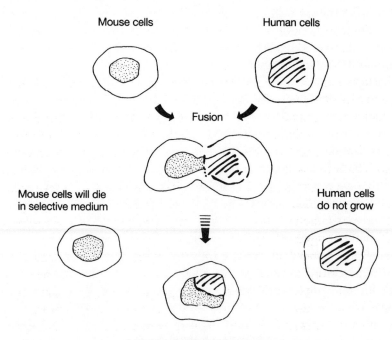

Hybrid cell with mainly mouse
chromosomes and human X will grow

human chromosome in the hybrid, the assignment of a marker to it will be immediate.

With the development of recombinant DNA techniques, any DNA sequence represented by a DNA clone could now be assigned to its chromosome, an approach first developed by Alec Jeffreys of DNA fingerprinting fame. Today it is even possible to identify fluorescence-tagged DNA *in situ* on a chromosome.

By 1970, the prospects for molecular genetics were only beginning to dawn on us, and I remember speculating about how it might be possible to develop human genetics with the same precision and refinement as was then possible with bacterial genetics. The outlook seemed enormously exhilarating even then, and that was before recombinant DNA technology, pioneered by Boyer and Cohen, had been developed. Even in our wildest dreams, though, we could not have predicted how far things would progress during the next quarter-century.

That year, I returned to England, to Oxford. The university's resources for carrying out somatic cell genetics and HLA work were good, and the academic environment was stimulating and friendly, but there was little or no molecular genetics. Our human gene mapping studies and work on the HLA system therefore flourished, but the first gene (an HLA gene) was not cloned in a laboratory of mine until after I moved to the Imperial Cancer Research Fund in London in 1979, and that was partly because my predecessor as director, Michael Stoker, had had the foresight to introduce recombinant DNA technology there.

My background – an apprenticeship in classical genetics under Sir Ronald Fisher, and experience in molecular genetics with Joshua Lederberg – proved to be a lucky one for applying DNA technology to the study of human disease, though the idea that genetic markers could be used to track inherited diseases in families, and so locate the corresponding genes on the chromosome, goes back to Haldane and Fisher. Haldane had, for example, written in 1927 that we might be able to use about fifty markers 'as landmarks for the study of such characters as musical ability, obesity and bad temper'. Given Haldane's famous irascibility, it is quite possible he was thinking of himself when he suggested this last attribute.

Haldane and Fisher had realized that variations then being defined in blood groups could be used in this way, and they pioneered statistical methods for studying linkage, using human pedigrees. In particular, Fisher appreciated that the use of genetic markers would not only aid the study of specific inherited diseases, but also help to define individual differences that do not follow simple patterns of Mendelian inheritance. Thus, in early 1932, he wrote that this 'work is going to lead to a greater advance, both theoretical and practical, in the problems of human genetics than can be expected from any further work on biometrical or genealogical lines.' He understood that the simple study of pedigrees and statistical analysis would not be enough and that precise studies were needed with well-defined genetic markers.

At the time, Haldane and Fisher's approach was frustrated by a lack of markers. Even with progress in blood group and enzyme work, there were still not enough markers to provide the density of information that is needed to map disease genes. But that situation was dramatically changed with the discovery of genetic variability that could be detected at DNA level.

The first simple and still widely used technique for detecting individual DNA differences came through the use of restriction enzymes and Southern blotting. This simple idea, as we discussed in Chapter Five, is based on the fact that if individuals differ in their DNA sequence at a site cut by a particular restriction enzyme, then this can be revealed because the DNA for one individual would be sliced at this position by the enzyme while another person's would not be. The difference could then be shown up in a Southern blot.

By 1977 it was realized that such markers could be used, for example, for prenatal diagnosis of a haemoglobin abnormality. Until then, the tracking of disease genes through families had been more or less hopeless because of the small number of normal genetic marker differences that were available. Chances of finding a marker near a disease gene on a chromosome were therefore minimal. But the advent of DNA polymorphisms completely changed the picture. Suddenly, we had an unlimited range of marker differences between individuals which we could study.

The enormous significance for the future of human genetics quickly became obvious to me. In early 1979 I wrote a letter with my colleague, Ellen Solomon, to the *Lancet* commenting on the use of DNA markers for studying the sickle cell gene. The letter ended with a paragraph emphasizing the wider application of these markers. 'Such a set of genetic markers could revolutionize our ability to study the genetic determination of complex attributes, and to follow the inheritance of traits that are so far difficult or impossible to study at the cellular level.' Now, for the first time, one could envisage constructing a complete genetic map of man.

Our letter originally included a more expanded discussion of how one might obtain DNA markers and use somatic cell genetics to map their chromosome positions. About a year later an American group led by David Botstein published a more extended account of this idea, stressing the obvious point that it is important that the DNA markers be as informative as possible and emphasizing the approach to constructing a genetic map using such markers in human families. Their paper had the flavour of new converts who had suddenly realized that DNA differences, which they had been studying as molecular biologists in the test-tube, could actually be used to study variation in families.

The crucial point is that the farsighted ideas of Fisher and Haldane were about to be realized. So let us examine why the genetic map is so important and how it can be applied to the study of human disease. First consider sickle cell anaemia, which we encountered in Chapter Three. It demonstrates the classical model. First the disease was discovered to be inherited in families following simple Mendelian rules; then the protein defect, the abnormal oxygen-carrying haemoglobin, was identified; and from this the responsible genetic lesion was found in the haemoglobin gene. The gene was found from an understanding of the chemical nature of the disease, in other words. But for many, if not most, inherited diseases that approach has not worked. It is just too difficult to sort out all the chemical processes of our tissues and organs.

Take cystic fibrosis. Affected children are sometimes first diagnosed because of their frequent lung infections. However, they may also have recurrent attacks of diarrhoea, while their pancreas can get blocked by sticky secretions. The clearest diagnostic feature of cystic fibrosis, however, is that their sweat contains unusually high concentrations of salt.

Over the decades, all sorts of ideas were put forward to explain the underlying defect for these various symptoms. Perhaps victims lacked an enzyme that would normally break down the sticky secretion, or was it something to do with the secretion itself? For years, at meeting after meeting, researchers claimed they had found the answer. They were always proved wrong, and in the end the secret of cystic fibrosis came from a totally different direction, one that completely ignored all knowledge of the nature of the disease – other than its genetics.

The first step, naturally, was to locate the gene; and using DNA markers, it was eventually pinpointed to chromosome 7. Once a gene has been localized to within a region of a few million base pairs, the challenge becomes one of narrowing down that localization still further. Finally, you have to pick likely candidate genes from those for normal proteins which lie within this reduced region. Each of these candidates is tested individually to see whether it fits the bill. In the case of cystic fibrosis that meant finding a gene which had an abnormality, the deletion of a few bases, that correlated strictly with the disease in many families and clearly explained the disease in the majority of cases. Thus, the gene was identified by the fact that it was mutated in affected individuals and not in normal individuals. Knowing about the disease had nothing to with identifying the nature of the defective protein. But having found the gene, profound clues were now available about the basis of the disease. The basic defect clearly has something to do with the movement of chemicals in and out of cells, a discovery that now promises to lead to totally radical treatments for cystic fibrosis.

This approach is sometimes known as 'positional cloning', since it relies on a gene's position in order to work out the basic protein defect. Achieving this has depended on the development of genetic engineering technology which now, at its most sophisticated level, enables large stretches of DNA to be analysed using clones that are a million bases in length, compared with those of only a few thousand which were typical of the technology's early days.

The first disease genes found through positional cloning were discovered in 1986. They were chronic granulomatous disease, an inherited disorder in which the blood's scavenger cells fail to gobble up and kill bacteria, and Duchenne muscular dystrophy. The inherited cancer of the eye, retinoblastoma, was discovered shortly afterwards, and three years later the gene for cystic fibrosis was identified. Now new disease genes are being found almost daily. These include genes for other inherited cancers, such as the bowel cancer susceptibility polyposis; the extraordinary fragile-X syndrome; and at least two genes for the premature ageing condition, Alzheimer's disease.

In each case the gene has been cloned purely from a knowledge of its position on a chromosome. And to do that, you need good luck mixed with a pretty hefty dose of sound laboratory management. The luck is required in finding a rare abnormal chromosome with a piece missing, a deletion that narrows down the gene's position, as was the case for polyposis. Alternatively, one requires a translocation between chromosomes, such as the Philadelphia chromosome created from an exchange of pieces between chromosomes 9 and 22, as we discussed in Chapter Six. These events help reduce the number of candidate genes that have to be tested in the search to find the real McCoy. As for good management, you need that for the hard slog as one guides researchers along a few million base pairs, uncovering genes on the way, and testing them for appropriate mutations. However, if we had already completed the Book of Man, then you would know what was there and would simply have to test a few candidates, without having to do the hard work of finding them in the first place.

The underlying genetic basis of tumours, discussed in Chapter Six, means that gene mapping has become a particularly important tool in cancer research. That is why the Imperial Cancer Research Fund has put so much effort into it. For instance, there are the losses of gene function. The genetic events leading to this 'knock-out' leave clues in the cancer cells which distinguish them from any surrounding normal tissue. For example, a cancer cell can lose part of a chromosome carrying the normal gene. When it does so, it will also lose all the other genes on that bit of the chromosome.

Suppose, then, that we have DNA markers for a chromosome, say chromosome 5 for a bowel cancer. Now an individual has two versions of the chromosome: one inherited from his mother and the other from his father. But if the cancer has knocked out one bit of one chromosome 5, then there will only be one full version left. By using markers that distinguish the paternal and maternal chromosomes, you can then ask which markers are lost on which chromosome. Those markers therefore help to pinpoint the position of the gene that is going wrong, and can be applied whether there are familial cases of the cancer or not. Indeed, that is how the role of the p53 gene was first identified in bowel cancers, though we now know it is mutated in at least half of all human cancers.

Cancer is, of course, a major killer. As you pass through middle age, you begin to notice the obituaries of your contemporaries. And the chances are high that they will have died of cancer, or heart disease. In Chapter Six, we looked at the genetic roots of the former; however, it is the latter which is responsible for nearly 50 per cent of all deaths in the West and it now appears that genetic factors play a major role in these – perhaps even more than the environment. Heart disease is, of course, complex and involves components such as increased blood pressure, or hypertension, and also atherosclerosis, a disease in which fatty plaques develop on the inner walls of arteries, eventually blocking them. As with cancer, however, there are some rare examples of clearcut inherited susceptibilities.

Consider the case of James Harrison who had a heart attack when he was only forty-five. He had atherosclerosis and his blood cholesterol level was unusually high. Naturally James was worried, particularly as his brother had died of a heart attack in his mid-thirties, suggesting that an inherited predisposition was involved. It turned out that James carried the gene for familial hypercholesterolaemia. This is a simple, inherited trait that leads to the build-up of very high blood levels of cholesterol, and this in turn is associated with the early onset of atherosclerosis and a 50 per cent risk of developing ischaemic heart disease. The disease is caused by a defect in cell receptors which take up lipids called low density lipoproteins (LDL). Their accumulation leads to blocking of arteries and to heart disease. About one in 500 individuals has this abnormality, a relatively high rate for an inherited disease, although familial hypercholesterolaemia still only makes a small contribution to the total numbers of heart disease victims. Nevertheless, as was the case with cancer families, an inherited condition has provided an important clue to the causation of heart disease.

And then there is high blood pressure, which has long been thought to run in families. Indeed, some estimates suggest that up to 80 per cent of differences in people's blood pressure could have a genetic basis. This is also reflected in the close parallels between blood pressures of identical, as opposed to non-identical twins.

This familial trend also figures in rats and mice, in which strains with very high blood pressure can be produced by breeding generations from individual animals with the highest pressure, a type of selective breeding that cattle breeders employ for improving their milk and beef yields. And when the high blood pressure rats were crossed with rats with low blood pressure and their offspring cross-bred among themselves, an intriguing fact emerged. Blood pressure was linked to renin levels.

Renin is released by the kidneys into the blood, where it acts on another protein called angiotensinogen. This reaction produces a second chemical called angiotensin, which in turn reacts with another enzyme to produce a substance which causes blood vessels to constrict and so increase blood pressure. It is a complex chain reaction, and genetic variation in any one of its stages is likely to cause blood pressure variation. We can therefore see why rat blood pressure differences seem to correlate with their renin levels.

Obviously, humans cannot be crossed at will, which means their genotypes cannot be studied for their links with blood pressure. Nevertheless, there are approaches that can be made, along the lines of the studies which linked HLA tissue types with auto-immune diseases. In Chapter Seven we saw that within a family, individuals who have ankylosing spondylitis also share the B27 HLA type. So if we focus our attention just on affected individuals in a family, they should share the genetic constitution that gives rise to their susceptibility. Thus, if we suspect a disease like high blood pressure has a genetic basis, then we can search for genetic markers that, in families, associate with individuals who have the disease. This linkage analysis, or genetic mapping, tells us whether our DNA marker lies close to a real genetic difference responsible for the disease. And if the genetic marker is very close to the disease gene then it may associate with the disease in the population and not just in families. The crux of the matter is that, given enough genetic markers, it becomes possible to sort out genetic components of complex attributes, not just single-gene ailments.

When this approach was used on a series of families, each of which had two or more brothers or sisters with clearly defined high blood pressure, the gene for yet another component of the renin-angiotensin system, the protein on which renin itself acts (called angiotensinogen) seemed to be involved. In all, these factors imply that one day we will be able to identify the specific difference in the angiotensinogen gene, and use that to identify people who are at serious risk of high blood pressure. Then these people can be targeted for diet or drug treatments to reduce this risk.

Valuable information about the location of human disease genes can also be obtained from studies of animals. In particular, there is considerable similarity between the human and the mouse gene maps. So if we know where on a mouse chromosome a disease gene lies, then we can soon locate its corresponding position on the human map. It is a very useful procedure. For example, a gene that affects the number of polyps or precancerous growths produced by the mouse equivalent of the human polyposis mutation has been mapped to a chromosomal region that corresponds to the tip of the short arm of human chromosome 1, which immediately suggests that some genetic susceptibility to bowel cancer may be associated in humans with genetic variations there.

However, we should not forget that the gene mapping approaches used to study high blood pressure or other diseases can also be applied to normal differences between people. In other words, positional cloning could be used to identify the genes for baldness or eye colour, or differences in powers to taste and smell. Already genes have been cloned for a complex set of smell receptor genes that seem to vary enormously between people, just like HLA types. One day we may even be able to associate this genetic variation with individual food preferences!

But perhaps the most intriguing type of variation between humans concerns our one true method of recognizing each other – our faces. Quite simply, faces are fascinating, and recognition of their features must be a deep-seated capability, having evolved most probably in our primate 'days'. Its importance can be gauged from the fact that there is a special cranial region devoted to it. We know this because some unfortunate people – some of those brain-damaged in car crashes, for example – can no longer use this region and therefore cannot identify even their closest family members by their faces. Instead, they have to recognize their wives and children by their clothes or way of speaking. This precise power of facial recognition probably evolved because of a critical need to distinguish one's own family, and fellow primates, from threatening foreigners, and must have predated the development of spoken language.

As for faces themselves, they are almost entirely genetically determined, as we can see from the startling similarity of identical twins' appearances. But whatever genes are involved in coding for facial features, they must be an enormous number to judge from the variety of faces we can see every day. No two faces ever really look alike, apart from those of identical twins. These genes may range over the entire genome, or they could cluster in a few complex groups, like those that code for HLA types. However, they must be shaped by different forces from those that normally act on human attributes. Facial similarity within a population must have been created over the aeons because we like to find mates who look a bit like ourselves, or perhaps our mothers or fathers. Such 'assortative' mating will keep similar face genes together in a population, and is probably a major source of differences in racial appearances. This is an important point, for tracing face gene variations across Europe and other continents and correlating them with patterns of migration should add a new dimension to the gene geography we discussed in Chapter Nine.

One crucial clue that might help us in this search could come from studying chromosome abnormalities. These are often associated with unusual facial features. For example, Down's syndrome – usually caused by an extra copy of chromosome 21 – was originally called Mongolism because victims can have narrow, slanted eyes characteristic of oriental faces. This suggests that chromosome 21 may carry genes for some facial features, which of course have nothing to do with Down's syndrome itself.

Unravelling the secret of the human face will clearly be a complex business, but it is nothing compared to the task of teasing out the genetic secrets of intelligence and behaviour. In these cases, we face real problems because we really do not know how to measure intelligence or behaviour very satisfactorily, and we tend to think of them in terms of value judgements. Either someone is very intelligent or he or she is intellectually limited, or bad or good. However, there are clearly many dimensions to individual differences in behaviour and intellect, and there is no doubt that they must have significant genetic components. As we pointed out in Chapter One, it is hard to believe that Mozart's precocious genius was not inborn. It must have been written, from conception, in his genes. And as we have also stated, perfect pitch may be one of the letters that make up that script. Certainly, the trait does seem to run in families, and when we uncover the responsible gene's position and structure we can then find out how it confers its mysterious prowess.

Of course, to find such a gene it would be an enormous advantage to possess the complete Book of Man. Once again we could look through its pages for likely candidate genes and test them until one was found that correlated with perfect pitch. This is a point that I made in a lecture at the American Society of Human Genetics in 1980. 'The whole DNA sequence will eventually be known, and also its meaning will be understood,' I said. 'This knowledge will have profound implications for all aspects of human activities and endeavours and surely will, in the long run, contribute positively to the betterment of our society.' As far as I am aware, that was the first serious discussion of the implications of a complete analysis of the human genome.

But if we are going to unravel the entire human genome, we should first ask: exactly how many genes are there on it? Obviously, the fewer there are, the easier is the task ahead. Yet for all our recent progress in molecular genetics, the answer to that question remains elusive. For example if we take the total length of the human genome – about 3,000,000,000 base pairs – and divide this by the average size of a gene – about 1,000 base pairs – we get a figure of three million genes. But we know now that only a few percent of the genome actually codes for proteins. As a result, most estimates put the number at about 100,000. Even that figure may be too high by a factor of two, I believe. Furthermore, genes fall into related families, like the HLA types, or antibody genes. What really matters is not the number of different protein-determining genes but the number of different gene families, of which there are probably only about 5,000 to 10,000. The Book of Man may be less complex after all.

The history of the Human Genome Project can be traced back to the 1970s, when Victor McKusick and Frank Ruddle, another developer of somatic cell genetics, set up series of international workshops to update information on the human gene map. These meetings were the first international mapping collaborations and became the focus for the genome project. At that time gene mapping was considered to be an esoteric, generally rather useless activity, although Victor McKusick had already started collating his monumental catalogue of 'Mendelian inheritance in man', his list of all known human genetic diseases which was first published in 1966, so the idea of gene mapping was already in the air. But the final push for the Human Genome Project did not come from this community. It came instead from scientists with a background in molecular biology, researchers who suddenly saw the opportunity for a grand programme that might challenge, in scale, the great projects of physics – its accelerators, mighty telescopes and satellites.

One such molecular biologist was Robert Sinsheimer, head of the University of California at Santa Cruz. He wanted to promote an institute for his campus, a foundation which would put his university on the map. Sequencing the whole human genome seemed to fit the bill exactly. Sinsheimer started planning in late 1984, and brought together many luminaries from the world of molecular biology to discuss the idea. He stirred the pot and interested the US National Institutes of Health (NIH) and the major private medical research funders. However, none of them would commit money solely to Santa Cruz.

Elsewhere, similar ideas were stirring. Renato Dulbecco, who won the Nobel Prize for his part in showing how viruses can cancerously transform cells in culture, advocated sequencing the whole human genome in the journal *Science*. Independently, Charles Delisi, head of Health and Environmental Research at the US Department of Energy, decided that national laboratories which had been involved in cold war atomic energy research might be able to redeploy their resources on a major peaceful biological project in the *glasnost* years. Again emphasis was placed on generating maps, though the opportunity to exploit the laboratories' engineering and computing resources on large-scale sequencing doubtless attracted energy department officials. Delisi had a crucial advantage – he was a senior US civil servant and was able to influence his mentors to put money behind the scheme.

And there was the Howard Hughes Medical Institute, which had become interested in gene mapping in the early 1980s. One of its advisers, Charles Scriver, a distinguished Canadian medical geneticist, persuaded the institute to organise a meeting at its Florida headquarters in February 1986 to discuss a human genome project. Scriver as a geneticist was well aware of the importance of gene mapping and the need to develop the databases to accommodate the vast amount of information that would be generated by the project. I remember the meeting vividly. There was strong support for a systematic mapping effort and the question of setting up a co-ordinating group of scientists was raised. So the institute decided to organize a major forum in July 1986 in Washington to bring together all the different parties interested in a human genome project.

I chaired the meeting, which was attended by James Watson and Walter Gilbert, as well as various geneticists from the NIH, the Department of Energy and a sprinkling of other non-United States scientists, including Sydney Brenner. The discussion was lively and the notion, proposed by some, that the project should simply sequence DNA from one end of the genome to the other without emphasizing the genetic map, was roundly criticized by myself and others. With a map we could home in on the important features of the genome very quickly. If we just laboriously sequenced without directing our efforts, a great deal of valuable time would be wasted on inconsequential parts of the genome. The meeting was a major catalyst for the eventual development of the human genome project in the United States, though it was also tinged with a little antagonism between the Department of Energy and the NIH. The end result, however, was beneficial for it brought the two bodies together in a semblance of a joint programme.

By now there was open discussion about the human genome project in the scientific community and in the pages of *Nature*, *Science* and other journals. One important contribution was provided by a Cold Spring Harbor symposium in June 1986, entitled 'The Molecular Biology of Homo Sapiens'. This brought together most of the key players in the field, including some from the Soviet Union. I gave the opening address, and ended my talk by pointing out that even if the project cost $3 billion, this was a modest price, the equivalent of constructing a few Trident missiles, and only a fraction of the cost of other scientific projects then under discussion, such as putting men on Mars. 'It is my hope there will be a world-wide consensus to pursue this challenging, but achievable goal: the complete characterization of the human genome,' I said. 'A not unrealistic target would be the end of this century – project 2000.'

By this time, James Wyngaarden, director of the NIH, had become a total convert to the human genome project. A highly respected scientist, he also commanded considerable influence with US Congress and helped establish a significant budget for an American attack on the genome. (By 1992 the combined annual funding from the US DOE and the NIH was close to $200 million.) Then in October 1988 James Watson was appointed the director of human genome research at the NIH, thereby giving the project enormous prestige and political visibility.

In Britain, Sydney Brenner had become head of a molecular genetics unit supported by the Medical Research Council (MRC) devoted to genome research. Sydney and I had many discussions about how to promote the human genome project, both nationally and internationally. I was concerned that there was a great deal of pressure for a co-ordinated project in the US, and discussions in the European Community – but no real effort in Britain. I wrote a letter to the Royal Society in 1986 in which I expressed my concern and suggested that the Society should play a role in the project as an objective, highly respected learned organization. I emphasized that the project was not 'simpleminded sequencing of the human genome DNA from beginning to end' and added that 'the proportion of functional DNA is sufficiently small that to sequence it would not be a major problem.' However, the Royal Society preferred not to get involved, arguing that it did not want to become identified with the promotion of one particular subject.

Fortunately, opportunities arose elsewhere. At an Imperial Cancer Research Fund reception, a cabinet office official asked me what key project the government Advisory Committee on Science and Technology should be considering. My answer was immediate – the human genome project. As a result, Sydney Brenner prepared a brief for this committee and he and I presented it. Sydney and I also proposed that the MRC should have a separately funded programme on the human genome, and this was accepted by the government. Eventually a co-ordinating group involving representatives from the MRC, the medical research charities and industry was set up under the chairmanship of the chief scientist in the cabinet office. Now many other countries have established genome programmes following the lead of the US and UK. In particular, the European Community has set up an active programme.

Of course, constructing a human genetic map depends on the availability of large groups of families as well as highly informative DNA markers, as we pointed out in Chapter One. And the ideal collaboration for establishing a genetic map is for researchers to pool their resources and analyse the same sets of families. It was to this end that Jean Dausset, the French pioneer of the HLA system, set up the Centre d'Etude du Polymorphisme Humain (CEPH). Its aim was to seek out highly informative families which could be used by different research groups working with various markers to create comprehensive gene maps. Dausset's first wife had been an art dealer, and through this interest he had been left a legacy of an extremely valuable collection of modern French paintings. When auctioned, these provided the money to establish CEPH – which shows, if nothing else, that it is possible for the arts and sciences to have the occasional fruitful collaboration.

However, although there was extensive communication between gene mappers, there was no guarantee the human genome project would be carried out on a truly international collaborative basis. Indeed, the high political visibility of the project in the US, Europe and Japan meant the direct involvement of government agencies of these countries and an element of unhelpful national competitiveness began to creep in. As a result, the idea of establishing an international group of scientists to supervise human genome activities was proposed in April 1988 at a meeting at the Cold Spring Harbor Laboratories. Victor McKusick was a key proponent of the idea, along with Sydney Brenner, who suggested the name Hugo (Human Genome Organisation).

In 1990 I started my three-year office as Hugo's second president. (Victor McKusick was the first.) Hugo's principal aim is to foster international collaboration in genome mapping and sequencing though it has had a difficult period of gestation. Perhaps too much was expected of an organization that had to be international and had to find financial support and capture the commitment of the scientific community before it could take on major programmes and convince governments of its usefulness. It was certainly not an easy task creating a body to act as a link between scientists and governments, while simultaneously ensuring international collaboration within the Human Genome Project with a minimum of unnecessary competition. Within a few years, Hugo acquired more than 500 members from more than thirty countries and developed a series of essential research programmes. The Howard Hughes Medical Institute, the Wellcome Trust in the UK, and the Imperial Cancer Research Fund have provided finance while the European Community has used Hugo to help co-ordinate research activities.

But if we have a Human Genome Project, this begs one essential question. Whose genome should be selected? It is the commonest question I was asked in the project's early days. In fact, it does not matter whose genome is picked. Eventually, what will be done will be the sequencing of several different genomes. But what does matter is to use information from the genes that are discovered to study further the nature of human differences – which is, after all, the underlying rationale for the project. How else can one discover disease genes or clusters which give rise to susceptibility to heart disease or mental disease?

A perfect example of this kind of study is provided by Luca Cavalli-Sforza who proposed setting up a major study of populations, particularly races threatened with being wiped out by the spread of Western civilization. Using a wide range of DNA markers, it should be possible to establish much more precisely the patterns of relationship between different human populations, as we discussed in Chapter Nine. This project is one of Hugo's major commitments.

And then, of course, there is the issue of medicine. The Human Genome Project will become the mainstay of the pharmaceutical industry in the next century because it will provide the basic information for targeting new drugs and other treatments. However, to benefit from these discoveries, the active commercial exploitation of discoveries will also have to be established. And hand-in-hand with this must come protection of intellectual property rights – in other words, patenting. This poses a problem in striking the best balance between protection, while at the same time promoting free exchange of information and materials between scientists. The aim should be to control commercialization without inhibiting research, though there may have to be a few constraints imposed on collaboration.

This difficult issue was brought sharply into focus in 1991 when the US NIH filed four patents for fragments of gene sequences that had been obtained from copy DNA that had, in turn, been derived from messenger RNA using standard techniques. By sequencing arbitrary fragments of cDNA, or copy DNA, data can be used to discover new genes and so work out their functions. However, attempting to patent these arbitrary gene fragments without any clear utility attached to them seems to contradict the three main tenets of patentability – novelty, non-obviousness and commercial utility. So far, such applications have, fortunately, been rejected by the US Patent Office, though the issue has generated a great deal of adverse comment from scientists.

So at what point does it become appropriate to patent? I believe the best balance will be arrived at by protecting the use of a sequence without actually patenting the naturally occurring sequence itself. This should enable free exchange of DNA sequences and information concerning them but will still make it possible to obtain an appropriate level of protection that would allow effective commercialization of new discoveries. In other words, we should treat DNA as an international resource from which products can be generated. These may come in the form of DNA clones, messenger RNA or proteins, and could be patentable. The basic DNA sequence would not be, however.

Whatever else, though, we can see that the rate of progress towards producing the Book of Man has been extraordinary. There is now a powerful, intrinsic momentum behind the project, and there can be little doubt that the book will be of inestimable value when we eventually finish writing all its pages. However, we must accept that at the same time important ethical, legal and social questions will be generated and these are the subject of our final chapter.

13

The Slippery Slope

To Miranda Quarry, cystic fibrosis was not a complete bolt from the blue. She had learned of its grim effects from magazines and from 'London's Burning', one of her favourite TV soap operas which had contained scenes of a couple struggling to cope with an affected child. Nevertheless, Miranda had no idea that the disease might ever touch her own life – until she took part in a prenatal screening scheme in Edinburgh in May 1992.

In her late twenties, Miranda and her husband David had been planning their first child for some time, and when Miranda found she was pregnant, she was referred to a pioneering project run by Professor David Brock at the city's Western General Hospital. This enterprise was one of a series that had just been set up to establish the best method for detecting the two million cystic fibrosis gene carriers in the British population, a prospect that had just become feasible thanks to those developments in modern molecular biology that we discussed in Chapters Four and Five. A simple mouthwash was used to collect a few thousand epithelial cells from inside Miranda's cheeks, their DNA was isolated, and then, using PCR, clinicians attempted to create multiple copies of a piece of the cystic fibrosis gene to determine if it resided on one of her chromosome 7s.

'I knew about cystic fibrosis, but I didn't know that as many as one in twenty-five people in Britain are carriers,' says Miranda. 'Even then I just thought I would be one of the twenty-four.' She was not. The mouthwash revealed she carried the cystic fibrosis gene. The Quarrys' trials were beginning. 'I was being sick throughout the day, and I had been in a minor car accident. Then I was told I was a carrier.'

The revelation was a shock, but not an overwhelming one, recalls Miranda. She was just unlucky, one of the 1-in-25 brigade. The chances that

David was also a carrier remained remote – until a second test showed he, too, had the mutant gene. In a few days, the disease's comforting statistical remoteness had been blown away. As we saw in Chapter Five, there is a one-in-2,500 probability that a birth in the West will produce a cystic fibrosis baby. For the Quarrys, two quick swills of water into a beaker had turned that prospect into a very real, one-in-four chance. 'My reaction was one of total disbelief. After everything I had been through, all I could think was: why? why? I was heartbroken. I wept buckets.'

Suddenly, the Quarrys were facing some very hard choices. In particular, they had to make up their minds, quickly, what to do if the next stage, an amniocentesis test – a procedure in which foetal cells are cultured and then tested for a particular genetic or chromosomal ailment – showed that Miranda's unborn child was afflicted with cystic fibrosis. Miranda was now seventeen weeks pregnant and despite careful, supportive counselling, which is provided for all women found to be carriers, the strain was painful.

'David and I decided we would terminate the pregnancy if the amniocentesis test was positive. Yet we both kept thinking: aren't we being selfish? Couldn't we struggle on with a sick child? But, in the end, I thought what it might do to us. One of us might not be there. How could a single parent cope with an affected child?'

A few days later, the Quarrys' trauma disappeared as quickly as it materialized. Amniocentesis revealed that the child had inherited no cystic fibrosis genes at all. The baby was not even a carrier. Miranda and David could settle down again in happy anticipation of the birth they had both wanted so badly.

The Quarrys' story is an important one, for it shows that the benefits brought with the maturing of molecular biology also carry ethical and moral 'side effects'. Some hard thinking, painful decisions and stress will be a direct consequence of the implementation of its techniques. And in the case of cystic fibrosis, the most significant inherited ailment in the West, there are several profound issues to be resolved – counselling, education of the public, and the application of resources, for example.

Consider the question of resources. Was prenatal screening really the best way to pick out carriers like the Quarrys? Miranda was, after all, well into her pregnancy when she went to Professor Brock's clinic. Surely, if both she and David had known their status, possibly even before they had met, a lot of heartache might have been avoided. Professor Brock agrees. 'Community testing – that is screening the entire country for carriers – is the preferable goal. It is the more expensive option, of course, because you will be testing many people who will never have children. It will also take longer to establish, so we are likely to face problems like Miranda's for some time.

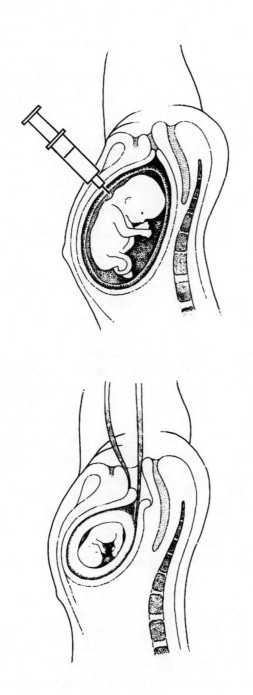

Screening the unborn. Chorionic villus sampling (left): a tiny amount of the chorionic villi, which surround the foetus, is removed, and its DNA tested. With amniocentesis (right), a needle is inserted through the amniotic sac that surrounds the foetus. Fluid is removed, and cells that have been sloughed from the foetus are isolated and tested.

In a sense we can look at efforts to try to stop cystic fibrosis as being like a game of cricket. We want to get a full quota of fielders to catch the ball – the cystic fibrosis gene – but the first in place should be the wicket keeper, in the form of prenatal screening.'

This latter form of testing is based on taking a sample of material from an unborn child. There are two principal ways to do this. Firstly there is amniocentesis, which was used to show that Miranda Quarry's baby was free of cystic fibrosis, and which entails a doctor inserting a needle though the amniotic sac that surrounds the foetus. A sample of fluid is then removed. Cells which have sloughed off the foetus into this liquid can be isolated and cultured before being given a desired DNA test. The other form of testing is known as chorionic villus sampling. In this technique, a tube is inserted into the cervix, and then into the womb, or through the abdomen. A tiny amount of the chorionic villi, which surround the foetus like the petals of a flower, is then sucked out. This material is of the same genotype as the developing child and it can be used as the basis for a DNA test.

Typically, amniocentesis has been carried out late in pregnancy, about sixteen weeks after conception. Chorionic villus sampling is usually carried out at nine weeks, by contrast. This has given it an important advantage. At this stage, an abortion is relatively simple and involves only a brief visit to a hospital. More importantly, it causes far less trauma to a mother. With amniocentesis, a woman often has not received the results of her test until she has reached the twentieth week of pregnancy, by which time she is carrying a fairly developed human being, and will have to go through induced labour; though this has become less of a problem with recent improvements which now allow doctors to carry amniocentesis sampling around the eleven-week stage in pregnancy, so reducing trauma and ameliorating prenatal screening stress. We should note, however, that both CVS and amniocentesis are associated with spontaneous miscarriages. For the latter, the figure is about 0.5 to 1 per cent. For chorionic villus sampling, it is about 1.5 per cent. Both factors therefore have to be taken into careful consideration before undergoing prenatal screening.

In any case, will it ever be replaced by community screening? And what would happen if it was? Will some people not misunderstand the concept of being designated a carrier? And is there not a danger that some stigma will cling to them? The answers to the last two questions will, regrettably, be affirmative unless we are very careful. An example of how things can go wrong is provided by screening that was established to pinpoint carriers of sickle cell anaemia in the United States in the 1970s. This was botched because programmes were set up without thought about what actions could be taken when carriers were detected. Some states made screening mandatory – but for black people only. Some, such as Massachusetts, actually declared that carriers of the sickle cell trait had a disease in its own right. The result was widespread discrimination over jobs, marriage and health insurance.

With proper counselling and improved biological education these problems could be avoided, however. We need only look at two highly successful programmes for screening – Jewish people for Tay Sachs disease (which we encountered in Chapter Nine) and London's Cypriot community for thalassaemia, diseases to which each group respectively is prone. These are now established, well-run, popular programmes even though both faced the prospect of upsetting sensitivities because of their associations with racial minorities. Counselling and education were vital in achieving success.

Providing these services on national scales for cystic fibrosis will require substantial amounts of public funding, and particularly education, for at present the average citizen lacks the background in biology and genetics to understand the implications of having carrier status – though we should be equally certain that that person will comprehend its significance once the importance of his or her condition is explained. Some counsellors believe the best time to do this will be during adolescence when young minds are mature enough to grasp the ramifications of carrier diagnosis, but still have time to consider its implications before having children. If we want to do that, however, we will have to change our ideas about education. At present, schoolchildren often give up science and become specialists in other fields at the age of fourteen and never think about the subject again. Molecular biology has got too close to the hearts of people's lives for that to happen any more.

In view of such considerations, the investment, and the careful explanations, some people might ask if genetic testing is really worth all the fuss. Do we want genetic testing that badly? These are reasonable questions, though we should be sure of our answers. Screening – both in terms of money invested by the state, and in terms of demand and uptake – is, unequivocally, a benefit to society.

Let us consider the first point – the investment – and look at the example of the Edinburgh prenatal screening project. In its first phase of operation, about 3,500 pregnant women were screened and 123 were found to be carriers. Of these, four (including Miranda Quarry) were also discovered to have carrier partners. And of this quartet, only one was then shown to have an affected foetus. (That couple chose to have the pregnancy terminated.) It may seem a like a lot of effort, screening 3,500 pregnant women to find only one affected embryo. Yet the whole process was still cost effective. The Edinburgh project was funded through an £80,000 grant from Britain's Cystic Fibrosis Research Trust. In comparison, the average cost of keeping alive one cystic fibrosis patient is more than £125,000. Such calculations sound brutal, of course. Nevertheless, they show that screening offers manifest value for money.

But there is a far better justification for introducing screening for genetic diseases. People want it. And a perfect example of this desire is provided by one of the very first diseases for which such a service was offered – thalassaemia, the severe and, until very recently, fatal form of inherited anaemia which we encountered in our discussion of gene geography in Chapter Nine. Before the advent of screening, parents who had discovered their carrier status with the birth of their first affected child were left in a terrible predicament. 'The only way couples could avoid having affected children was to stop having children at all,' says Dr Bernadette Modell of University College Hospital, London, one of the pioneers of thalassaemia screening. 'In poor Catholic countries, where there were no contraceptives available, that meant husbands and wives could no longer even sleep with each other. They could scarcely cuddle each other for fear it would "go too far".'

With the advent of genetic screening for thalassaemia that position was transformed. Not only carrier couples, but communities throughout the Mediterranean – Sardinia, mainland Italy, Greece and Cyprus, as well as immigrant communities in Britain – took up the service enthusiastically. And when affected foetuses were found, offers of termination were usually accepted, even in Sardinia and Italy which have predominantly Catholic populations. As a result, numbers of births of babies with thalassaemia have plummeted. In Sardinia, for example, they declined by 70 per cent in a decade, which shows that the urge to have healthy children is an extremely powerful one. And what is true of thalassaemia will also be true of cystic fibrosis one day. Once decisions about education, type of screening and resources have been established, we can expect an equally enthusiastic response from Caucasian communities who, as we saw, are the ones principally at risk of the disease.

One approach that has been proposed for halting births of children affected with genetic diseases would be to take the process of avoidance back one step, by trying to discourage matings between carriers in the first place. According to this idea, people could be screened to see if they carried a gene for sickle cell, cystic fibrosis or another ailment. Then they would be encouraged to advertise the fact so that they could avoid socializing with, and occasionally marrying, other carriers of the same disease gene. Not surprisingly, the idea is unpopular, for it is fraught with problems. We saw from the bungled introduction of sickle cell testing programmes in the US that there is a great danger that people might confuse the concept of being a carrier with an actual diagnosis of an illness. By emphasizing a person's carrier status, we could inadvertently brand someone with a genetic stigma. Apart from the prejudice this might invoke, members of such a biological 'under-class' might be drawn together, instead of being persuaded to stay apart, and that, of course, would only lead to an increase in affected pregnancies.

And then there are genetic ailments which raise special problems of their own, such as Huntington's chorea and Duchenne muscular dystrophy, con-ditions which, with cystic fibrosis, formed the core of Chapter Five. Let us look at Duchenne, first of all. As we saw, effective screening is now possible for families at risk of the disease. But as a third of all cases are caused by ran-dom, spontaneous mutations of the X chromosomes throughout the general population, we can never hope to eradicate the disease merely through pre-natal tests of Duchenne families. Of course, we could test the foetus of every pregnant woman for Duchenne mutations and so screen the whole popula-tion. The trouble is that the techniques involved – amniocentesis and chorionic villus sampling – both pose slight risks of triggering miscarriages. As we have seen, the chance of a pregnancy being terminated accidentally in this way is of the order of 1 per cent. And given that Duchenne children account for only 0.02 per cent of births, we would be ruining large numbers of pregnancies just to find one Duchenne birth.

Then we come to the question of Huntington's chorea. Because a carrier of its deadly gene is automatically a victim, it has also posed awkward, sometimes intractable questions. Of course for many people the emergence of tests has been a godsend. Individuals have found they are free of the dis-ease in whose shadow they have lived most of their lives. And even in cases where a positive test result was given, confirmation has sometimes been viewed as preferable to constant uncertainty. As one woman told Nancy Wexler: 'God, get it over. I'm so tired of wondering.' In addition, there is the prospect of avoiding giving birth to affected children. As a result, surveys suggest that two out of three individuals at risk of Huntington's would be prepared to take the test. Equally, that statistic suggests that one in three would say no, while other surveys indicate this fraction may be even larger.

The trouble is that there may be strong insistence that people in this lat-ter group take a test even though the disease's unpleasant symptoms would not manifest themselves for many years. For instance, would a medical school want to train a physician as a neurosurgeon if it was shown he or she had the Huntington gene, since the early stages of the disease are character-ized by tremors and irrational behaviour? And would the military want to train a person with the gene? The answer to these questions is probably no, which means stress for the unfortunate individual who is being coerced into taking a test that he or she would prefer having nothing to do with. But there are also family pressures to ponder. Either husbands or wives will want at-risk spouses to take the test to see if their future offspring will be affected, or older children might decide to take the test themselves as they contemplate their own marriages. In these latter cases, a positive result for a son or daughter would obviously reveal that the at-risk parent was also affected. Resolving these problems will be extremely difficult.

Most vexed of all is the question of abortion. To those who oppose termination of foetuses as an issue of principle, the development of genetic screening tests causes particular anxiety. They believe we may stop more and more pregnancies for increasingly irrelevant reasons. 'What might now be seen as a noble and altruistic wish to prevent people from leading "useless" lives, may easily take us to the day when self-preservation rules, when we belong to a society that does not want to waste its time or money on people who are going to be a burden on health services,' says Mrs Nuala Scarisbrick, of the anti-abortion organization Life. 'That is the philosophy of elitism and the master race.'

In other words, where will it stop? If we terminate foetuses known to have cystic fibrosis or Duchenne muscular dystrophy will we create 'a search and destroy' mentality that leads to abortions for more trivial defects – diabetes, left-handedness, ankylosing spondylitis or colour-blindness, for example? Are we on a slippery slope to moral degeneracy? It is an easy allegation to make, of course, though it is based on not a single shred of evidence. For a start, those of us who live in Western societies exist in an ethos in which some women choose to have abortions in order not to interfere with their careers, for example. It therefore seems unfair to accuse parents who elect to terminate a forthcoming child with blighted health, after careful, painful debate of dragging society down to some new moral nadir. Nevertheless, constraints on some 'genetic' terminations would seem to be reasonable, particularly those concerned with screening tests for sex selection, for example. In some societies there is a strong wish to have a boy as a firstborn child, and selection following sex determination after amniocentesis or CVS is clearly possible. But should it be allowed? In the West, most people believe the answer is an unequivocal no, for apart from using abortion for what amounts to be trivial social engineering, there could be a damaging distortion of the distribution of men and women, with disastrous social consequences.

In any case, screening for thalassaemia and sickle cell anaemia has been with us for almost a decade now, and there is not the faintest indication that it has brought us nearer to the mindless practice of eugenics, the science of improving the human condition through selective breeding. Indeed, most consequences of screening have been unequivocally positive, as Professor Peter Rowley of Rochester University has pointed out. 'Pre-natal diagnosis has had a "pro-life" effect for couples who previously avoided pregnancy

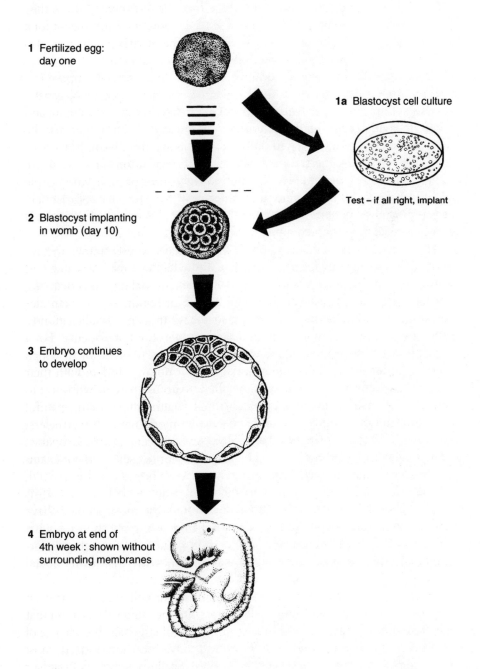

1 Fertilized egg: day one

1a Blastocyst cell culture

Test – if all right, implant

2 Blastocyst implanting in womb (day 10)

3 Embryo continues to develop

4 Embryo at end of 4th week : shown without surrounding membranes

Picking the right embryo. How blastocyst screening and implantation can be used to select embryos free from an inherited condition.

because of a genetic risk. Now they are willing to conceive. Furthermore some couples choose pre-natal diagnosis with no thought of termination but rather to prepare for the birth of a child with special needs.'

Not that abortion is the perfect option. It just happens to be the only effective method we have for dealing with serious inherited maladies at present. In future, this is unlikely to be the case, and some ingenious alternatives are already being developed. One consists of flushing out a blastocyst, the ball of cells that forms the very early embryo. This blastocyst can then be screened and reinserted only if genetic tests prove satisfactory, thus completely avoiding the trauma of abortion, though it does not avoid the moral issue of termination of a life. Alternatively, using in vitro fertilization (IVF) techniques, it is possible to create several embryos from sperm and eggs provided by parents, and implant only those that have been screened, and cleared of the inherited ailment carried by one or both of the parents.

These procedures elude the trauma of abortion and avoid accusations that one is killing an already formed individual, though critics still argue that IVF techniques involve unacceptable tampering with human individuality. This is a hard argument to sustain. The ball of cells formed by a fertilized egg before implantation has no real identity, so one cannot argue that something distinctive is lost each time a newly created embryo is discarded.

Just how far one should exploit this technology is a different matter, however. Consider the case of Anima Ayala, a Californian girl who developed leukaemia when she was seventeen. She had no brothers or sisters with the correct HLA match, and so a bone marrow transplant was impossible. In their desperation, Anima's parents chose to have another child. And this time, the HLA type of the new baby, Melissa, matched Anima's. Two years later, Melissa provided the bone marrow for a transplant that saved Anima's life. It is an intriguing story, because the Ayalas were fortunate in one sense – for Melissa's HLA type might not have corresponded with Anima's. As we saw in Chapter Seven, there is only a one-in-four chance of a child's HLA type matching that of a sibling. Using pre-implantation screening of embryos created by IVF, however, it would have been possible to assure a match. But is that an ethically correct procedure – to have an HLA typed child, discarding other embryos created in the process, only to save the life of another? Put that way, we might be tempted to say no. But if the child is wanted and loved and saves another's life, is there anything wrong with that? It is a rather awkward point.

If nothing else, we can see that IVF and pre-implantation screening are very promising technologies, though they are currently limited by the fact that there is a very low take-up rate (about 20 per cent) in the womb of embryos made this way – which means we can assume that screening and terminations will be with society for some time yet.

In fact, the underlying problem behind all the issues we have discussed is a simple one. Molecular biology has, in an astonishingly brief space of time, been able to pinpoint the sources of much of mankind's genetic misery (at least those woes associated with single-gene mutations), but it has yet to find the means to put them right. We know the causes, but cannot yet neutralize them. That means we are in an interim phase in modern genetics. We can spot the dangers, but we generally remain powerless to neutralize them, other than by carrying out terminations. However, that situation is already changing. The first tentative steps to remedying these problems, with the creation of new drugs, and of course genetic therapy, are already being taken.

Indeed, there are already examples where our powers to detect the perils to a newborn child are matched by our ability to put them right. One such condition is called phenylketonuria (PKU). Sufferers have an inherited flaw that causes them to be deficient in an enzyme responsible for turning the amino acid phenylalanine into another very similar amino acid called tyrosine. The result is a child whose body has high levels of phenylalanine and low levels of tyrosine. The accumulation of phenylalanine and its toxic by-products leads to a severe mental retardation if left unchecked. This imbalance also triggers a chain of biochemical events that cause children to be deficient in melanin pigment, and this gives them their characteristic fair complexion and blond hair.

PKU can be detected very simply, however. At British and American hospitals it is now standard practice to take a tiny pinprick of blood from a newly born baby which is spotted on to filter paper and given a straightforward test. In this way, PKU victims can be detected. But the crucial point is that something can be done about such a diagnosis. By giving a child a diet that contains little phenylalanine, the brain damage with which it is associated can be prevented. It is as simple as that, though there is one problem to watch out for. If a woman has PKU and is not maintained on the special diet while she is pregnant, then mental retardation will be induced in her foetus by the high levels of phenylalanine in her blood.

And then there is the issue of cancer-risk diagnosis. As we saw in Chapter Six, it is possible to identify individuals with inherited predispositions. For such people, screening procedures such as breast mammography or searching for pre-cancerous growths in the colon have special benefit. The development of the tumours to which they are disposed can be spotted well in advance, and as we know, early detection of a cancer means a greatly improved chance of cure.

One example is provided by mutations in the oncogene p53 which are found in about half of all human cancers. Very occasionally, a p53 mutation occurs in a sperm or egg cell and is inherited through the germline. This gives rise to 'cancer families' whose members carry the p53 mutation in all their cells and who are therefore in danger of succumbing to a variety of cancers, particularly those of the bone, when young, and breast, when adults. On the other hand, even with this bleak genetic disadvantage, about 20 to 30 per cent of these p53 carriers will avoid cancer throughout their lives. The situation is further confused because it is not easy to screen for some cancers, such as those of the bone, to which a family may be prone. Their diagnoses will therefore cause anxiety, which in some cases will be unnecessary, because they will not get cancer, or because doctors will not be able to screen for the particular tumour to which they will eventually succumb. We can therefore see that the issues are ambiguous, though there is no doubt that knowledge of their inherited predispositions will benefit many members of such a 'cancer family'.

In addition, the current predicaments of genetic screening should disappear once technology can put right the mutated culprits that it uncovers. In theory at least, abortions will no longer be necessary because we will be able to correct those life-threatening faulty genes. Such developments are unlikely to be with us for at least a decade, though the creation of the Book of Man, in the form of the Human Genome Project, will play an important part in hastening its arrival. Meanwhile society must accommodate genetic screening and its associated issues with consideration and patience. Its benefits far outweigh its deficits, after all.

In the meantime, we can expect to reap enormous dividends from the mapping, and then the sequencing, of all the genes that go together to make 'a piece of work' like man. As we saw in Chapter Five, we have already entered a new stage in our reading of the Book of Man. No longer will laboratories use personnel to carry out the laborious sequencing of a desired gene, as with cystic fibrosis and other diseases. Researchers will simply home in on a bit of chromosome, look up the pages of the Book of Man to find what genes lie in that region, and then select ones which might have a mutation that fits the symptoms of the ailment under study. Laboratories will not be filled with sequencers, but virologists, cell development specialists and all the other experts needed to exploit genes for their pharmaceutical potential.

And we should be very clear on this point. The future of the development of new pharmaceuticals in the twenty-first century will rest squarely with the mapping and sequencing of all our genes. With the identification of the triggers of disturbed protein pathways, we should be able to develop drugs that will block them, so halting the pernicious progress of diseases, for instance in conditions such as Alzheimer's disease. We should also be able to attack tumours with pinpoint precision, and correct the dreadful physical deterioration associated with immune disorders like rheumatoid arthritis.

Those will be the first, most striking, consequence of scientists' authorship of the Book of Man. And there is a very important subsidiary point that we should make about the impact of the grand march of molecular biology – for we should be very clear about the way in which these early benefits will have been achieved. The subjects of the Human Genome Project may be men and women, but the majority of critical insights were achieved through simple experiments – on animals.

This one of the most significant messages that should be taken away from this book. As readers will have noticed throughout these pages, there have been a considerable number of occasions when animals have been used to make crucial advances in molecular biology. Indeed, the very origins of our understanding of DNA's role in genetics came from Griffiths' experiments with the pneumococcus bacteria in mice, as we described in Chapter Three. In addition, our knowledge of blood groups, the immune system, transplant rejection, tumour viruses, the role of growth factors in cancer development, and the cause of diabetes were all achieved through experiments on animals. Nothing could more flatly refute the claims of animal activists and anti-vivisectionists who argue that experiments on animals, allegedly so different from humans, have done little for medical research, and are irrelevant for understanding people. It is one of the most pernicious lies of our time, as our book makes abundantly clear. For those who have closely followed the progress of medical research there can be no doubt that appropriately controlled animal experiments are ethically justifiable. Those who think otherwise are usually misguided or just incapable of accepting the fact that all progress has a price. A small remaining fraction simply argue that no advancement is worth the price, although they represent a very small percentage of the population.

Certainly, in the case of animal experiments, the price can be measured in the lives of animals (mice and rats) normally treated as vermin by society. Used sparingly, and within reason, it therefore does not seem a prohibitive expenditure to carry out such experiments in order to create life-saving drugs.

On the other hand, we can see that there will be consequences regarding some of the other developments of modern genetics. Earlier in this chapter, and also in Chapter Ten, we touched upon one of the thorniest of these – DNA confidentiality. Consider the following, as yet hypothetical, example. Suppose a p53 mutation is identified in a person as a by-product of another study, for example as part of a study on tissue typing. Should that individual be told? The detection of the mutation may be of considerable help in screening for a cancer that otherwise might have killed that person. However, there is only a finite risk of such a prognosis and a person could end up suffering anxiety about a disease to which he or she will never succumb.

What we should do with the individual genetic information that will be obtained through the Human Genome Project, and controlling its use, are some of the key issues that will have to faced in the next few years. The natural starting point is to assume absolute confidentiality, and the right of privacy with respect to knowledge of one's own genetic make-up. Unfortunately there are sufficient exceptions to this rule to show this should not be considered a hard and fast law. It may be beneficial to an individual, for example, to reveal to an employer an inherited tendency to develop allergic reactions to certain substances, so that these can be avoided in the workplace. Equally, though, employment screening carries the clear risk of generating discrimination and, perhaps, of lessening the imperative to maintain a 'clean' environment.

In short, data protection is a thorny issue. It treads the fine line between the rights of an individual not to have information used to their disadvantage, and the value that such knowledge may have for an organization such as a charity for fundraising, or government for improving its health care.

Consider the proposal to set up databases of DNA fingerprints, which was touched upon in Chapter Ten. It would undoubtedly help catch criminals, for any biological trace found at the scene of a crime could be checked with this central register and a clear lead provided for police. Similarly, a DNA database of soldiers would help identify those killed in action when other forms of identification are impossible. These are straightforward, beneficial applications. Yet it is unlikely that there would be any public consensus to implement them, because of the old slippery slope argument. Civil liberties groups argue that if we start assembling DNA databases, it is a very short step to adding other information to this compilation until we have created vast ledgers of electronic data about the citizens of our nations. Central files bulging with details about every British or American or French person would be open to abuse, they say. Certainly, until proper safeguards can be proposed which might satisfy such fears, it is unlikely such databases will introduced.

The detailed analysis of inherited susceptibilities also interests life and health insurance companies. By outlining individual genetic predispositions, we should, one day, be able to calculate life premiums, and those without proclivities to serious illness might expect to pay reduced premiums as a result. After all, there are specially reduced rates for non-smokers and non-drinkers. So does this mean that individuals will one day have their premiums based on genetic read-outs of their disease propensities? The answer is, probably not, for the principles of insurance are based on spreading risk. If everyone's chances of succumbing to illness could be matched precisely to a premium, there would be no business left for the life insurance assessor. In any case, calculating individual risks is likely to remain an almost impossible business, despite the development of gene 'print-outs'.

Private health insurance is a different matter. In such a situation, premiums must take into account the existence of well defined severe genetic disease. But the premiums could then become exorbitant. So some government safety net will have to be devised, like Britain's National Health Service, which spreads the risk uniformly without reference to individual susceptibility, so covering the needs of the child with cystic fibrosis and the high blood pressure victim. As the medicine of the future depends more and more on genetic information and targeting preventive measures as a function of an individual's genetic make-up, it seems hard to envisage that health care systems of the future will not, in one sense or another, be in the form of a national health service.

Such a service raises another question: should screening always be voluntary? Should parents be allowed to refuse tests for a disease like PKU? These are difficult questions to answer, for they touch on the vexed issue of trading the rights of the individual against the broad needs of society. There are precedents, however. Many countries in the West have passed laws which require people to wear seat-belts in cars or which ban the advertising of tobacco products. Such legislation covers the same broad ground, though for similar laws to be passed to make some genetic tests compulsory there would need to be widespread public debate.

These are arguments that pertain to the present and they are, in a sense, practical issues. However, it is the more problematic, uncertain future that worries most critics of the new genetics. They believe gene therapy today may lead to something much more terrible tomorrow. Genetically altering a human being is the unacceptable face of molecular biology, they argue. Such critics see a terrifying Brave New World of designer-made human beings, populating some biologically determined hell. Scientists could end up by playing God, tinkering with pedigrees and perhaps unwittingly, or even deliberately, creating Frankenstein monsters (though these critics never explain why researchers would ever want to concoct such monstrosities in the first place).

Of course, the application of any new idea or technology in medicine inevitably poses ethical, social and legal questions. But the problem seems to be particularly acute for genetics, perhaps because our recently acquired understanding of the processes of inheritance has unravelled one of the great mysteries of life, striking at the very heart of our self-identification as human beings. People look at the progress made since Crick and Watson discovered DNA's structure in 1953 and wonder when scientists will start to tamper with life in barely understood unpredictable ways. But is this really the future for modern genetics?

Well, in Chapter Eleven we saw that gene therapy will initially concentrate only on somatic cells, on tissue that is unconnected with the business of reproduction. For example, a patient's bone marrow cells could have a defect corrected in the laboratory and then be put back in their body to provide a missing enzyme. Or a cancer patient could have tumour cells removed to be modified with genes for growth factors in such a way that when re-inoculated, they stimulate the patient's immune system to destroy the cancer. These techniques are sophisticated and technically demanding, but they do not differ in principle from other forms of disease treatment. They do not in any way affect a person's inheritance, and raise no significant new ethical questions.

This point was stressed in an editorial in the *Lancet* in January 1989. 'Why does gene therapy seem to raise major ethical issues when only somatic cells are involved? It cannot be because new genetic material is being introduced into the patient since organ and bone marrow transplants do just that routinely. Moreover, treatments with radiation and certain drugs will cause alterations in the genetic material. Neither can it be the safety argument since there is no reason to believe that gene therapy will be more dangerous than a host of other treatments. All new medical interventions carry risks and there are well established procedures for introducing, for example, new drugs. It is hard to see how anyone could object to curing disabling genetic diseases.'

Indeed, it would be unethical not to implement gene therapy for patients with life-threatening ailments, points out French Anderson, one of its pioneers. 'Patients with serious genetic diseases have little other hope at present for alleviation of their medical problems. Arguments that genetic engineering might someday be misused do not justify the needless perpetuation of human suffering caused by unnecessarily delaying this potentially powerful therapeutic procedure.'

The genetic modification of sperm or egg cells is a different matter, however. To begin to think seriously of such experiments suggests we might perhaps be moving towards a tomorrow of designer humans, though removing an obvious, fatal mutant from the human gene pool would still be unequivocally beneficial for mankind. Extracting the Huntington's chorea gene from the biological legacy of a family cannot be viewed as anything but a good idea, for instance. This form of gene therapy is known as negative germline therapy, because it involves the removal of unwanted genes, and it does not particularly worry observers. It is the positive version of germline therapy – the addition of attributes – that sends the real shivers down their spines.

At issue is the extent to which we would use this technology to improve intelligence or modify inherited make-ups in desirable but unpredictable ways. Could we end up with genetic supermarkets where parents could select the characteristics for a forthcoming child? And what would they select when they got there – the features of an angel, perhaps? This latter notion was once addressed by J.B.S. Haldane. 'To create an angel would require new mutations, since we appear to have neither the genes for wings nor for moral excellence,' he pointed out. In other words, there is no absolute achievable goal of biological excellence, despite the claims of fanatical racists like Hitler.

In future, though, could some malicious dictator try to create a controllable band of slaves implanted with genes that engender compliance? Ideas like this frighten some people into thinking molecular biology could lead to a rebirth of eugenics and fascism. But why should there be any connection? Hitler did not need any sophisticated knowledge of genetics to perpetrate his extraordinary evil, after all. In fact, the unfortunate taint of racism with which genetics has been linked is the very reverse of the true position, as we saw in Chapter Nine. Far from underlining the concept of distinct racial differences, modern molecular biology has shown that it is almost non-existent. As we have seen, individual variation within any population is far greater than average differences between populations, even between major racial groups. Studies also tell us there is no sharp boundary between populations, especially those that are geographically closest together – and which most often battle with each other, as in the former nation of Yugoslavia. Ironically these are the populations that are the least distinguishable by any objective genetic criteria.

At present, modifying the germline might seem ridiculously unlikely. But if we had asked a doctor twenty years ago if it were likely that we might soon use genetically engineered viral vectors to alter the DNA of a victim of a lethal inherited disorder, we would probably have been laughed at. The fact that we can now carry out such treatments shows we should never underestimate the rapidity of the progress of science.

But if we develop the capacity to carry out positive germline therapy does that automatically mean we will exploit that ability? To many critics, the answer is a clear affirmative. Once a technology becomes feasible, it will inevitably be put into practice, they argue. This is the 'slippery slope' argument again. If something is possible, it will be implemented. The cracking of the atom 'inevitably' led to the creation of nuclear bombs, it is alleged. This is untrue. The construction of nuclear bombs occurred for fairly logical, military reasons that had a lot to do with the progress of the Second World War. It is a precarious argument to assume every scientific development will always be turned to dangerous use. Such an attitude will only stifle the beneficial obtaining of new knowledge, the very process by which Homo sapiens has stamped his mark on the world. Instead of ignoring scientific wisdom we must, as a society, cultivate the maturity not to misuse it. The alternative is to paralyse all progress because we are afraid some individuals might distort its potential – scarcely the sign of a culture that has any confidence in itself.

And in any case, when debating germline therapy, would it really be so bad if we added genes for height to small people, or for hair to the bald, or good eyesight to the myopic? Probably not. But if we did, would it then be deemed unacceptable to insert intelligence or athleticism genes? It probably would, for the very notion of the sanctity of human individuality would be too grossly offended. Just where we get off the slippery slope is therefore a matter for society to choose, as the US Institute of Medicine and National Academy of Sciences made clear in its report on human gene therapy. 'The prospect of parents turning to enhancement techniques to produce perfect children raises important questions about the value of the individual in our society and about the appropriate use of limited health care resources.' Fortunately, given the timescales that are involved in our progress towards such an eventuality, we have plenty of time to debate the issues and resolve them.

In that interim we need a calm appraisal of genetic engineering. The term is convenient, but at the same time is emotive and confusing. In the public's mind, genetic engineering muddles all the different applications of recombinant DNA technology that we have described in this book – for example, for tackling inherited disease, understanding cancer, gene therapy, and the genetic manipulation of sperm and egg. And mixed in with this genetic potpourri comes in vitro fertilization, people-cloning and, no doubt, some hidden notion of the creation of Frankenstein's monster. The beneficial applications of recombinant DNA technology are therefore mixed with highly speculative concerns.

It is a point emphasized by the *Lancet* editorial quoted earlier in this chapter. 'It is curious how frightened people are of genetic engineering compared with, say, child abuse. Genetic engineering has so far damaged no one. By contrast, smoking, Aids, drugs and alcohol have caused massive damage to children in utero. What one needs is an educated public. They need to be sufficiently DNA literate.'

DNA literacy: that is precisely what we should be inculcating in society, and it is hoped that this book will play a part. Through the pages of the Book of Man, we have unfolded the drama of the unravelling of the composition, structure and role of DNA, Watson's 'most golden of molecules'. It is a story of adventure and discovery, one that strikes at the essence of being a human. It is hard to see how one could get any closer to our own interests, for the question of our biological lineage touches, as we have seen, on understanding our role in evolution, and survival as a species. It also offers the prospect of creating new medicines, tracing our recent history and a host of other benefits. Far from worrying, we should relish the prospect before us. We are entering a golden era.

Acknowledgements

The authors and publisher would like to thank the following sources for pictures reproduced in the book:

Mic Rolph (line illustrations and diagrams);

RCAHMS (Ferniehirst Castle, p.2); Photo Researchers/Science Photo Library (Watson and Crick, p.39); Science Photo Library (Rosalind Franklin, p.40); James King-Holmes/Science Photo Library (Frederick Sanger, p.49); Peter Menzel/Science Photo Library (Walter Gilbert, p.49); the Wellcome Institute Library, London (Edward Jenner, p.110, and Robert FitzRoy, p.141); Gary Haley (map of Polynesia, p.167); Leicester University (Alec Jeffreys, p.176); the Press Association Ltd (Karen Price reconstruction, p.189);

and the Science Photo Library for the photographs reproduced in the colour section, in addition to the following individuals and organizations: Francis Leroy, Biocosmos (plate 1); Biology Media (plate 2); Professors P.M. Motta and S. Correr (plate 3); Philippe Plailly (plate 4, top and bottom left); James King-Holmes/Cellmark Diagnostics (plate 4, bottom right).

Index

Note: References in italics indicate illustrations.

249